Fifty Years of CITRUS

THE FLORIDA CITRUS EXCHANGE : 1909 - 1959

JAMES T. HOPKINS

UNIVERSITY OF FLORIDA PRESS
GAINESVILLE • 1960

A UNIVERSITY OF FLORIDA PRESS BOOK

Library of Congress Catalog Card No. 6010227

Printed by the Pepper Printing Company
Gainesville, Florida

DR. J. H. ROSS
President, Florida Citrus Exchange, 1913-1924

Preface

THIS IS THE HISTORY of the Florida Citrus Exchange. It encompasses in general detail an outline of the most significant developments of this organization during the fifty-year period from 1909 to 1959.

As the first federated marketing cooperative in the Florida citrus industry, the Exchange has served as the voice of the relatively large group of citrus growers who market their fruit through the Exchange system. It has been in this capacity that the Exchange has grown to prominence and considerable influence within the industry as well as on markets both domestic and foreign.

It will be disappointing to some that this history cannot hope to record in even briefest detail the names and contributions of many individuals who have served either the Exchange or the industry with considerable note during the past half-century. Neither is it possible within the limitations of this publication to include specific mention or historical note of the many associations that are charter members of the Exchange, or of the many other associations and affiliate shippers that make up the membership as its exists today.

Because this is the history of an organization, it will deal with individual accomplishments only in the light of their effect on the course of progress of the Florida Citrus Exchange. It must be noted here, however, that there exists the most interesting prospect that some future historical endeavor will one day take up the individual accomplishments of the industry's early leaders who, through strength and sheer determination, fashioned a mold for the future progress of the Florida citrus industry.

The extent to which the Florida Citrus Exchange has influenced the industry during the past half-century must be obvious at once to the unbiased mind. While appraisal of the use of this influence may vary to a great degree, there can be little doubt that the Florida Citrus Exchange has always functioned in the interest of its members, who make up a large percentage of the total citrus grower population of the state. This is true, in its most uncomplicated analysis, because the Florida Citrus Exchange was organized by growers in their own interest. Its exclusive purpose through all these years has been the marketing of its members' fruit on a cooperative basis, returning to those growers the highest possible margin of profit for their harvests. Policy and control of the Exchange is and has always been placed specifically in the hands of growers who elect to market their fruit through the many Exchange associations and associate shippers throughout the Florida citrus belt.

It seems quite natural that the early history of the Exchange should be synonymous with the recent history of the Florida citrus industry, and the reader without specific interest in the record of this cooperative should, nonetheless, find the organization and development of the organization interesting because of its parallel to the development of the industry itself.

With regard to the industry, this history concludes that problems and seasonal difficulties have occurred with such frequency that one is forced to the certain observation that no season or period of seasons has ever been free from trouble of one nature or another. The Florida citrus industry has never, it seems, known a normal season, nor for that matter has it ever found a workable pattern to apply to the long-range solution of each season's particular problems on the basis of prior experience with a similar problem.

Conversely, the bulk of today's problems are in themselves old and familiar. Indeed, a great many of them existed even before the half-century began. Progressive production and new marketing techniques have complicated these problems in many cases, but the difficulties of the citrus farmer of 1909 are not vastly different in basis from those of the grower of 1959.

And, like the problems, many of today's corrective endeavors are similar to those of yesteryear. There is, for example, little difference in the search for expanded consumer demand through advertising. Much of the Exchange's national advertising of the 1920's is strikingly similar in copy, idea, and art, to the advertising used by the industry today. Testimonials by outstanding sports figures, educational copy with regard to vitamins, citrus recipes, and the extravagant use of Florida's sunshine theme were all used extensively in that era by the Exchange in its advertising program, just as they are used in today's industry-wide advertising campaigns.

The historic similarity of problems, ideas, and methods of doing business within the industry during the past fifty years carries over into virtually every phase of the Florida citrus industry. To those who find interest in this history, it is probable that this fact will do no more than remind them that there is truth in the well-worn phrase that there is nothing ever completely new in Florida citrus.

The preparation of this history represents many months of research in the yellowed, musty, and timeworn records that form the documentary library of the Florida Citrus Exchange. Its accuracy is based on these official documents, and this history is presented as an actual record of the progress of the organization.

Those long hours of research and composition involved in this undertaking have been in themselves particularly rewarding to the author, for they have provided a new and searching scrutiny of this intriguing industry which captures men with the deftness of an uncompromising maiden and thereafter refuses to set them free.

J. T. Hopkins

Tampa, Florida
March, 1960

Contents

Under the beneficent influences of a semi-tropical sun, fanned by the breezes from three seas, results the finest climate on earth summer and winter, especially adapted to citrus culture. Away from the maddening confusion of modern manufacturing, commerce and transportation, the citrus industry invites men of refinement and industry to a gentleman's occupation which brings one so near to the great white throne in a Florida orange grove that, with the soul attuned, he can hear the angels sing.

—J. H. Ross, M.D., 1922.

CHAPTER 1

Introduction

IT WAS a particularly hot and humid day. For two days conventioners had been making their way into Tampa on horseback, by wagon—and a few by auto and train. And, while the convention hall itself was the flourishing Tampa Bay Casino, conventioners spread out over the town, lending a feeling of nervous excitement to townspeople thereabout. Everywhere the talk was about citrus, for this was the annual convention of the citrus growers of Florida. The year was 1909 and the month was June; and it was on June 1 that the group convened at 10:00 o'clock in the morning.

For some thirty years the Orange Growers Convention had met in the summertime to discuss problems and techniques of all phases of the Florida citrus industry. But in later years one problem had begun to emerge that overshadowed all others in every discussion: Production of Florida citrus had reached nearly 6,000,000 boxes—enough, it was said, to necessitate some method of curtailment in order to curb oversaturation of the nation's markets. Others felt that the answer was more probably to be found in some kind of organization that could control marketing procedures to such an extent that demand could be kept high, and prices likewise at a high level. From this latter group had come a proposal four months earlier, on February 26, that a group of intelligent citrus growers and businessmen visit the California citrus industry to study the obviously successful business methods employed on the west coast. The original proposal would have sent a committee of six men to make the study. The committee had grown to forty-six people before the trip got under way, all of them traveling at their own expense.

All this was in the past, on June 1, 1909. The Committee of Forty-Six had returned and were prepared to report their impressions to the assembly of the Florida Orange Growers Convention. None of the more than a hundred conventioners knew exactly where the day's activities would lead, but there was an air of expectancy—something important was in the making. Fans fluttered as conventioners took their seats in the Casino auditorium. Here and there were ladies, but the great majority of the assemblage were shirt-sleeved, sun-tanned citrus farmers intensely concerned with the business at hand.

At 10:00 o'clock sharp, Josiah Varn of Bradenton, secretary of the convention, brought down his gavel and waited patiently for the crowd to settle.

Preliminary matters came first, including a communication from the Peninsular Telephone Company extending the free use of its lines to the members of the convention who desired to call their homes during the two-day meeting.

Josiah Varn tendered his resignation as secretary of the convention, saying that the pressure of personal affairs had become such that he could no longer adequately care for the affairs of the convention. His resignation was accepted on a prevailing motion. Charles Walker was appointed to replace Varn as secretary.

After a score of relatively minor matters were disposed of, Dr. F. W. Inman of Florence Villa took the rostrum. Intelligent and highly respected in the industry, Dr. Inman's name was to appear frequently in connection with the proceedings of the convention. "On the 26th of February," he began, "we convened in convention to view the better method for harvesting and marketing crops of citrus fruit. Previous to that, I had made a trip to California for the purpose of studying conditions and their method of conducting the fruit business. I made the statement at the convention that, if we could send a delegation of intelligent orange growers and business men to California to study their methods, I was well convinced that they would return and bring news to their fellow growers which might aid them in forming a substantial organization for the growers of our state."

Dr. Inman went on to say that the committee had gone to California and had returned, and that they were ready to report their findings to the convention. Subcommittee chairmen had been appointed for the purpose of expediency, and one by one these chairmen filed up to the rostrum.

Eugene Holtsinger of Polk County came first. Then came Thomas Palmer of Tampa, W. S. Hart of Hawks Park, T. B. Quinby, and L. R. Skinner. Their reports covered grading, legislation, and transportation.

It was at this point that the subcommittee chairman, Thomas Palmer, obviously irritated at the slowness with which the meeting was proceeding, took the floor to deliver an important message. "I arose for the purpose of suggesting somewhat of a break in these proceedings at this point," he said. "I will preface my motion with this remark, that I think so far as the general purpose of the California Exchange is concerned, that is understood, and minor matters and details can be considered later. Now it seems to me we are here for the purpose of either adopting or rejecting the California plan. We know what the status is; to carry it out, there will have to be a lot of committee work done. We want to get right down to work, and the balance of the reports from the different committees can be attended to after this question is settled."

With this message concluded, Palmer addressed Dr. Inman, suggesting that he call for a motion adopting the California Citrus Exchange plan by the Florida group. A general discussion followed.

A delegate rose to ask if the railroad development in California had not been largely responsible for the progress of the California Fruit Exchange. Dr. Inman promptly answered the question: "The railroads did not bring about this Exchange, gentlemen; it was the one big stick that did it, the California Fruit Exchange. We have a condition not inferior to California's. We have just as good timber as California; there is no advantage they have that we have not. I want to call your attention to all the results and success of the California Fruit Exchange. Without it, southern California would be scarcely on the map."

Another delegate rose to his feet and asked: "How can we successfully organize an exchange here without knowing what we are organizing? I understood it was the purpose of this meeting of the convention to act on the information that was received from the different committees. It seems to me that the hearing of the reports is absolutely necessary before we can make an organization."

Thomas Palmer replied that most of the reports yet to be heard were mechanical in nature and would lend nothing to any decision as to whether or not the group should organize. He added: "The main thing to do is to say 'We are going to organize.' It seems to me the proper thing to do is to either adopt or reject the California plan. It is before you now. So far as the Exchange is concerned, you cannot know any more about it than you know now by listening to the different reports. Remember, there is work to be done. There are by-laws to be prepared, and we must arrange the charter."

Palmer then looked directly at the audience and challenged, "Right now, either adopt or reject the California plan. I move that this report be accepted and a committee appointed to get to work at once." Despite the objections of several delegates, Palmer's motion was seconded by J. W. Carson of Frostproof and was finally carried unanimously by the group.

A nominating committee to name a general committee on organization was appointed by Palmer. The committee was to include two members from the east coast, two from the west coast, and two from the central orange-growing section of Florida. These committeemen were Thomas Palmer of Tampa, Josiah Varn of Bradenton, W. S. Hart of Hawks Park, F. A. Lane of Fort Myers, Finias E. Parker of Arcadia, and L. W. Tilden of Oakland.

Later that day, the nominating committee brought in the names of thirty members of the convention who would form the committee for organization. This list included Dr. F. W. Inman, Eugene Holtsinger, and J. W. Sample for Polk County; Thomas Palmer, W. E. Heathcote, and Solon Pemberton for Hillsborough County; W. H. Cook for Putnam County; G. W. Lainheart for Palm Beach County; Josiah Varn and John B. Singletary for Manatee County; Finias E. Parker, Terrel Fielder, and W. H. Hooker for DeSoto County; D. S. Bourland, G. T. Raymond, and H. E. Heitman for Lee County; J. F. Corrigan for Pasco County; L. W. Tilden, W. C. Temple, and J. H. Lee for Orange County; O. W. Sadler and E. S. Burleigh for Lake County; F. G. Sampson and W. B. Gray for Marion County; W. A. Fulton for Hernando County; W. S. Hart and H. B. Stevens for Volusia County; E. P. Porcher and H. S. Williams for Brevard County; Thomas McCarty for St. Lucie County; M. S. Burbank for Dade County; and L. H. Montgomery for Alachua County.

With the appointment of the committee for organization, Dr. Inman called for the floor and said: "I take the liberty to impress upon this new committee their duties. We realize that this is a great measure for the State of Florida. We will meet the most complicated conditions that can possibly exist. Remember that we are going to meet opposition. We are not thoroughly qualified, nor are we educated up to the work in hand, but we can improve every day.

"The leopard cannot change his spots in a single night, and the work that is before you cannot be consumated at once. I am an old man, and not able to do much, but we have young men among us who can go ahead and develop. Take hold young men, and make this organization one that we will all be proud of. We can bring our fruit from the grower to the consumer, as does the California organization, and we can bring a better quality of fruit."

Thus on June 1, 1909, the first few faltering steps toward the organization of the Florida Citrus Exchange were taken.

Although there were other matters to come before the convention, the ones mentioned formed the basis upon which was created the Florida Citrus Exchange, which for many years to come would rule as the largest single element in the Florida citrus industry.

CHAPTER 2

1909-1910

NO SOONER had the Florida Orange Growers decided to organize into a marketing organization, than financial matters began to press upon them. There arose the long-range need for assessment against growers for support of the new organization, but more urgent was the need for funds for the present. E. O. Painter of Jacksonville, taking note of the urgency for immediate money, made the first contribution, saying: "I am willing to contribute $100 to start the ball rolling."

W. R. Fuller of Tampa then commented: "I will make my contribution $250 if Mr. Painter will."

Whereupon E. O. Painter raised his contribution to $250.

Other initial contributors included Overstreet Crate Company of Orlando; Wilson and Toomer of Jacksonville; W. W. Clark, mayor of Bartow; W. C. Temple of Winter Park; W. A. Merryday Company of Palatka; Swann and Holtsinger of Tampa; C. M. Broadwater of Orlando; Carvey Brothers of Fort Myers; W. H. Baum of Crystal River; Lee County Packing Company of Fort Myers; Ferguson and Long of Winter Haven; George E. Coplin of Winter Haven; C. M. Barton of Jacksonville; Connor and Shallaberger of Bartow; N. H. Fogg of Altemonte Springs; F. D. Waite of Palmetto; H. A. Ward of Winter Park; Miles and Lane of Olga; W. D. Taylor of Ocala; E. L. Pearce of Clearwater; a Mr. O'Brien of Seffner; and F. H. Adams of Dunedin.

With the temporary arrangement for money completed, the convention broke up into a series of committee meetings that occurred almost daily for the next six weeks. But, spurred on by the determination of the organizational leaders, the committees accomplished many things by July 22, 1909, when the entire membership of the new organization was called into convention.

Progress of the various committees was reported by Chairman Thomas Palmer in a lengthy address. Palmer told the group that the Florida Citrus Exchange was, as of that date, "a living active being in the State of Florida." He added: "It is here for all time to stay, for the benefit of the growers of Florida; it is here under the control of the growers of Florida, and will continue under their control."

Palmer reported that committees had selected the city of Tampa for the permanent headquarters of the Exchange, and that the businessmen of the city had agreed to pay the Exchange's office rental for two years. He told the grower audience that the board of directors had elected Dr. F. W. Inman as president of the Florida Citrus Exchange, and that W. B. Gray of Tampa had been elected first vice-president. He also reported that R. P. Burton, a former Florida citrus man who had gone to California after the freeze of 1887, had been released by the California Fruit Exchange to the Florida Citrus Exchange as its first general sales manager.

Other important developments up to the meeting of July 22 included the agreement with California that the new organization could "borrow" J. A. Reid, California Fruit Exchange organizer, for the purpose of organizing local associations which would build and support their own packinghouses. A business manager, M. E. Gillett of Tampa, had been named, and Thomas Palmer was named as the Exchange's first attorney.

It is interesting to note that Palmer made the first pledge of fruit for marketing under the new organization, but at the meeting of July 22 all members were urged to sign contracts pledging their fruit to the Florida Citrus Exchange. It was at this meeting that Palmer resigned from the board of directors to accept employment as the Exchange's legal counsel and R. C. Peacock was named first cashier of the Exchange.

Total pledges of fruit now reached 201,600 boxes, pledged by W. C. Temple, R. P. Burton, Thomas Palmer, G. T. Raymond, J. W. Sample, M. S. Burbank, W. B. Gray, Solon Pemberton, M. E. Gillett, G. W. Lainheart, Josiah Varn, H. E. Heitman, W. S. Hart, H. E.

Carlton, D. S. Bourland, O. W. Sadler, E. S. Burleigh, J. F. Corrigan, L. W. Tilden, W. E. Heathcote, J. E. Kilgore, W. A. Fulton, W. M. Pierson, C. A. Boynton, J. B. Pyland, F. H. Adams, J. B. Singletary, G. E. Koplin, and Z. C. Chambliss.

From this original pledge of fruit the Florida Citrus Exchange gained stature quickly, attained financial stability through borrowing privileges made possible by a $50,000 line of credit, and became in itself a legally certified cooperative organization for the purpose of marketing Florida citrus fruit.

The minutes of the first few meetings of the Exchange are lengthy and, as board meetings still do, rambled from one subject to another before entirely disposing of the first. Perhaps the most critical problem facing the cooperative was the organization of local associations for the purpose of banding growers together and packing their fruit.

Under the direction of J. A. Reid, the California organizer, local associations were in various stages of organization into sub-exchange districts. By August 25, 1909, one association to be known as the Haines City Growers Association was already in the process of packing fruit under the Exchange system.

On September 25, 1909, General Sales Manager Burton told his board that he had appointed salesmen in Detroit, Montreal, Toronto, New York, Buffalo, Rochester, Albany, Syracuse, Pittsburgh, Cleveland, Toledo, Evansville, Fort Worth, Memphis, Selma, Hattiesburg and Meridian. Also in September, 1909, the Florida Citrus Exchange hired its first press agent, R. H. Rose, who was employed for a period of three months at a salary of $50 per week and expenses. Employed as assistant to the general manager was R. C. Peacock, who had previously served the Exchange as cashier.

At the October, 1909, meeting of the board, it was determined for the first time that the Exchange could expect to handle in excess of one million boxes of fruit during its first year. An estimate of operating expenses placed the cost of operating the Exchange at $243,000. The first retain was established at 12 cents per box, based on the expected movement of two million boxes of fruit during the second season. For the balance of 1909 the transactions of the board of directors apparently centered mostly on money matters. At periods during this time, their balance sheets indicated an extremely precarious position, and, individual members of the board often used personal funds to meet payrolls and other operational expenses.

Some bitterness apparently existed between the board and various Tampa banking facilities at the time, because of the banks' reluctance to accept contracts for fruit as security for operational loans to the Exchange. More than one member of the board suggested that all money transactions be transferred to other cities in the state, but as the newly organized cooperative grew in size and stature, these motions were quieted.

It is interesting to note that an appropriation of $5,000 was made in this era for advertising purposes, and represents, so far as this writer is able to determine, the first Florida citrus consumer advertising program.

Several problems involving nonpayment of assessments by associations had already begun to appear, and a move to deduct these assessments from sales returns was eventually adopted early in the year 1910. Assessments at this time amounted to 12.5 cents per box for the central exchange plus 2.5 cents for the sub-exchanges.

While a complete list of associations in the exchange at the end of 1909 is not in the minutes of board meetings, reference is made to a number of sub-exchanges and associations, including Florence Villa and the Haines City group, Arcadia Citrus Sub-Exchange, Coconut Grove Citrus Growers Association, Crescent City Citrus Growers Association, Highland Citrus Sub-Exchange, Indian River Citrus Sub-Exchange, Lake Como Citrus Growers Association, Leesburg Citrus Sub-Exchange, Manatee Citrus Sub-Exchange, North DeSoto Citrus Sub-Exchange, Naranja Citrus Growers Association, Orange County Citrus Sub-Exchange, Owanita Citrus Growers Association, Orange City Citrus Growers Association, Pinellas Citrus Sub-Exchange, Polk County Citrus Sub-Exchange, Pomona Citrus Growers Association, and Volusia County Citrus Sub-Exchange.

By the end of 1909, the Northern sales offices payroll included thirty-nine employees and was reaching significant proportions. The payroll of the Tampa headquarters had likewise been growing at a rapid rate, so that it, too, included 39 employees by January 1, 1910.

These employees were: R. P. Burton, sales manager; M. E. Gillett, general manager; Charles G. Harness, cashier; O. G. Cook, traffic manager; W. C. Hewitt, inspector; R. L. Goodwin, organizer; R. H. Peacock, assistant to the general manager; Joel Whitley, assistant to the sales manager; L. Satterfield, stenographer; E. D.

Dow, claim clerk; A. S. Hall, bookkeeper; K. S. Clark, recorder; H. E. Long, diversions; G. H. Rehbaum, stenographer; D. E. Brunner, stenographer; Jessie Wauchope, stenographer; John Boyd, stenographer; Eugene Reppert, stenographer; C. M. Adams, Jacksonville locals; Anna Pexa, stenographer; Evelyn Owen, stenographer; A. L. Strickland, railroad records; C. J. Gunn, incoming telegrams; Robert Sinclair, incoming telegrams; B. J. Cobb file clerk; C. J. French, telegrams; Fred Parslow, telegrams; R. W. Erbaugh, stenographer; George E. Bulwinkle, outgoing telegrams; Sam Harrison, mail clerk; Rosa Cuthbert, stenographer; R. E. Norman, mail clerk; Frank Parker, mail clerk; Taylor Murphy, file clerk; Henry Torres, telephone; and George Pexa, messenger.

At the beginning of the new year, 1910, the Flordia Citrus Exchange had accumulated accounts receivable amounting to $25,468, furniture and fixtures valued at $5,633, and had $12,702 in cash. Liabilities amounted to $35,314. A brief note in the minutes with regard to this financial condition reads as follows: "The above covers assessment on fruit shipped up to January 15th. District Managers' salaries paid to January 15, and office salaries to February 1. The surplus will pay all brokerage and other bills. We are, therefore, ahead the amount of assessment on fruit shipped between January 15th and February 1, 1910, estimated at 100,000 boxes."

This, then, was the general picture of the Florida Citrus Exchange at the close of 1909. The board of directors, first appointed by the nominating committee in July, continued to meet weekly, and their discussions ranged widely over such subjects as count and packs to special springs for fruit wagons.

The board at this time included Dr. F. W. Inman, president; W. B. Gray, first vice-president; J. W. Sample, second vice-president; and E. Holtsinger, W. E. Heathcote, W. A. Fulton, W. S. Hart, H. E. Heitman, W. C. Temple, S. L. Griffin, and Josiah Varn. Thomas Palmer continued as the attorney for the Exchange.

CHAPTER 3

1910-1913

THE YEAR 1910 began with little change in the progressive structure of the Florida Citrus Exchange. The minutes for that year include reference to a meeting in which A. H. Kay, Exchange general Northern manager, brought a Mr. Fleming, president of the Georgia Peach Growers Association, to Tampa to discuss the possibility of the Exchange handling peaches through their sales organization during the following summer months. The Exchange board looked favorably on the proposition at that time, but took no positive action.

Also in April, attorney Thomas Palmer, who had been one of the stalwarts of the Exchange organization, asked for favorable consideration of his request for resignation as the legal counsel for the Exchange. His request was accepted and William Hunter was subsequently appointed. Thus began a long association with the Exchange—one that lasted until the present legal counsel replaced Hunter in 1934.

The resignation of M. E. Gillett as general manager was also accepted in 1910. Gillett, who several times had discussed this matter on a personal basis with members of the board, seemed overburdened by the many details of the office. In his formal request for resignation, he told the board that he was not at heart an office man, that he missed the out-of-doors, and that his health had suffered as a result of the confinement to the office. The board accepted his resignation and named W. C. Temple general manager.

Not long after the employment of Temple, a matter of relative rank arose between the general sales manager, R. P. Burton, and Temple, in which the board was asked to clarify the position of each. In a long letter to members of the board, Burton requested that they make it definitely clear to all employees and affiliates that the general manager had direct and general supervision over all departments and functions of the Exchange. This was accomplished to the credit of both Burton and Temple, and the end result was apparently a better organization with a clear-cut channel of staff authority.

The death of Dr. F. W. Inman in November, 1910, was a critical blow to members of the organization who had leaned heavily on the aging man. His death was deplored by many who anxiously viewed the future of the cooperative, and who were concerned about the mold the organization might take without him. Beloved and respected by the industry, Dr. Inman had carried much of the responsibility for convincing growers of the value of cooperative marketing. His passing moved the board to compose a long resolution in his memory. One paragraph of the resolution seems to sum up the sentiment of the members of the Exchange for their deceased president: "His integrity, his benevolence, his sterling character, stand as a monument to his memory, and a fit example for our emulation; and should nerve us to our greatest efforts to consummate and perpetuate the work he so well began."

Dr. Inman was buried at Lake Alfred on a small plot now known as Inman Park in recognition of the outstanding life he led. His final resting place overlooks the Citrus Experiment Station, where men of science today pursue dreams of greater things to come in citrus culture and process. To the researcher of these historic items, there is a nostalgic note in the death of Dr. Inman who, only a few short months earlier had said: "I am an old man, and not able to do much, but we have young men among us who can go ahead and develop. Take hold, young men, and make this organization one we will all be proud of."

Other commemorative resolutions flooded into the Tampa office from virtually every association in the Exchange system and from the California Fruit Exchange, as well as from industrial acquaintances of the highly regarded president. In the inexorable pattern of life, Dr. Inman's words were a prophecy of things to come, and the younger men were, indeed, destined to take hold. The minutes still include the short hand-written note of Dr. Inman's widow in response to the board's resolution.

The Exchange's first vice-president, Eugene Holtsinger of Polk County, stepped into the difficult role of top executive and, to his credit, the Exchange continued to develop under the careful guidance of the board of directors.

At the close of 1910, the board of directors included Eugene Holtsinger, W. B. Gray, J. W. Sample, W. E. Heathcote, George Koplin, H. J. Drane, W. N. Camp, L. W. Tilden, Josiah Varn, W. A. Fulton, J. M. Weeks, J. P. Mace, M. S. Burbank, W. S. Hart, and C. E. Stewart.

While 1909 and 1910 were years full of problems and almost daily crises for the Exchange's board, it is apparent that board members were conscientiously devoted to the work they had patterned for themselves. By early 1911, however, the routine of weekly board meetings and the generally favorable financial position of the Exchange had begun to make these meetings somewhat monotonous for many members. There were times when there were not even enough members to make up a quorum for the meetings. The situation became so drastic that a list of board members and their attendance at meetings was published in the February 1, 1911, copy of the official minutes, in the hope that this would help to alter the situation. The number of absences ranged from those of Eugene Holtsinger, who had missed only three meetings during the period from June 7, 1910, to January 25, 1911, to those of C. E. Stewart who had failed to attend twenty-seven meetings during the period.

The attendance record apparently did little good, for the minutes of February 22, 1911, are contained on one page, which bears in large letters the notation "Only Two Members Present. Absent . . . All The Rest Of The Directors. They Know Who They Are Without Mentioning Any Names. More Interest Should And Must Be Taken In Attending These Meetings."

On March 15, 1911, a meeting took place that laid the foundation for matters of signal importance. The minutes are quoted:

The general manager [W. C. Temple] presented to the board in detail certain correspondence exchanged by him with Mr. Robert Kennedy Duncan, director of Industrial Research and professor of Industrial Chemistry, at the University of Pittsburgh, and at the University of Kansas, in regard to the utilization of cull oranges, and of orange juice.

After full discussion, it was moved by Mr. Koplin, duly seconded by Mr. Drane, that a committee consisting of Mr. Holtsinger, Mr. Gray and Mr. Temple be, and hereby is, appointed to look into the matter of the work done and results accomplished by Mr. Duncan, and report to the board at a later meeting as to the advisability of having Mr. Duncan come to Tampa for a conference in regard to instituting a line of research in Florida as to the utilization of cull oranges and of orange juice.

The April 15 minutes of the board meeting further discussed this matter:

Mr. Temple representing the committee composed of himself, Mr. Holtsinger, Mr. Gray and Mr. Burton, which was appointed by the board at a previous meeting to take up with Professor Robert Kennedy Duncan, the matter of the preserving the juice and utilizing the by-products of the Florida orange, reported to the board that Professor Duncan had come to Tampa, upon their invitation, and had conferred with the committee.

And that the aforesaid committee would recommend to the board, the signing of a contract with the University of Pittsburgh, or University of Kansas, for an investigation to be conducted by Professor Duncan along the lines mentioned for a term of two years at the expenditure of $1,000 per year in support of the investigation, the amount and nature of the bonus to be paid the researcher to be decided upon at a later date.

On the basis of information from all known available records, this research represents the very first endeavor to preserve Florida citrus juices in such a manner as to create an additional utilization factor for the Florida citrus industry. The foresight of the board in this matter can well be recognized by today's citrus leaders, and it is, perhaps, the first in a long series of contributions by the Exchange to the benefit of the entire Florida citrus industry.

On May 9, 1911, Florida Citrus Exchange shipments neared the seasonal 1,000,000 box mark which had been predicted the year before, and receipts for the 1910-11 season for the first time went over the $100,000 level, totaling $106,393.69 for the period until May, 1911. At the conclusion of this season, the Exchange board included Eugene Holtsinger, W. B. Gray, J. W. Sample, George Koplin, W. E. Heathcote, W. A. Fulton, J. M. Weeks, Josiah Varn, W. N. Camp, H. J. Drane, J. P. Mace, L. W. Tilden, W. S. Hart, G. W. Holmes, W. T. Carter, M. S. Burbank, and C. E. Stewart.

The minutes of the first regular meeting of the newly elected board for the 1911-12 season show that most of the board members for the preceding season were carried over. The board continued with Eugene Holtsinger as president, W. B. Gray as first vice-president, and W. E. Heathcote as second vice-president. The directors were George E. Koplin, H. J. Drane, L. W. Tilden, Dr. J. H. Ross, W. S. Hart, J. P. Mace, W. T. Carter, Edward Parkinson, Barney Kilgore, Josiah Varn, W. A. Fulton, G. W. Holmes, and J. M. Weeks. Not long afterward, however, Barney Kilgore resigned as a director, and was replaced as the Pinellas Citrus Sub-Exchange's representative by Delisle Hagadorn. Another change in the directorship was to occur at mid-season with the death of W. N. Camp, who was replaced on the board by Jack Camp of Ocala.

The volume of Exchange fruit in the 1911-12 season was far below the hopes of the board, and on April 1, 1912, the Florence Villa Citrus Growers Association presented a resolution of confidence to the board of directors in which they expressed satisfaction for the method in which the Exchange had marketed their fruit, gave credit for the "satisfactory" condition of the industry to the Exchange, and pledged 60 per cent of all Polk County fruit into the Exchange.

The situation had become so dark by May 1, 1912, that the board prepared and passed the following resolution to the effect that unless greater interest was shown in the Exchange, the board would be forced to dissolve the entire organization:

Be it therefore resolved first, that the present board of directors will not take upon themselves the responsibility of effecting the dissolution of the Exchange; second, that it will leave final action in this matter to be taken by the new board of directors to be elected at the annual meeting of the Florida Citrus Exchange the first Tuesday in June; third, that it will recommend to the incoming board of directors that if by the first of June, 1912, an amount equal to 40 per cent of the fruit crop in the State of Florida is signed up with the Exchange, that the new board continue to operate the Florida Citrus Exchange for at least one additional year; fourth, the present board of directors of the Florida Citrus Exchange call to the attention of the incoming board that if the present organization is continued, that at the close of the present fiscal year of the Florida Citrus Exchange, to wit: August the first, 1912, there will be a shortage of $30,000 to be met, and they advise the present board was obliged to finance a similar deficiency at the closing of the previous fiscal year, and feels that it has done its full duty in this direction.

The membership at the third annual meeting on June 4, looked with some disfavor upon the resolution, however, and countered with action of its own:

Moved by Mr. Sample, seconded by Mr. Temple, that the retiring board of directors of the Florida Citrus Exchange turn the matters of financing and deficit over to the members of the new board of directors for their careful and prompt attention, without attempting to specify the details by which they shall handle these matters. Motion unanimously carried.

The newly elected officers of the board for the 1912-13 season included Eugene Holtsinger, president; W. B. Gray, first vice-president; and Dr. J. H. Ross, third vice-president. Members were L. W. Tilden, G. W. Holmes, J. P. Mace, H. E. Maury, Dr. W. C. Richardson, J. M. Weeks, W. T. Carter, G. W. Harp, Ed Parkinson, George E. Koplin, W. E. Heathcote, J. W. Sample, W. A. Fulton, and J. R. Williams.

Some problems involving loyal and efficient packinghouse managers arose during the summer of this season, and the board of directors unanimously agreed to employ seven packinghouse managers from California for the associations requiring their services. Nowhere in the minutes, however, are the names of these transplanted Californians listed. It is reported, however, that this procurement was actually accomplished and that second-generation Californians are still at work in the industry.

Assessments for the 1912-13 season were established by the board at its meeting on October 2, 1912, at 17½ cents per box, an extra 2½ cents being added over the prior year for advertising purposes. This advertising fund was destined to grow with the years, but the first major step in national advertising came during the October 16, 1912, meeting, when the board approved expenditures of $12,000 for newspaper advertising in New York and the New England states, with an additional $3,000 for an advertising campaign in the *Good Housekeeping* magazine.

It was during 1912 that the Exchange moved its central offices in Tampa to new and much finer quarters in the Citizens Bank and Trust Company building in the heart of the city. A lease was arranged for occupancy of most of the eighth floor of the building, but before being executed, it was changed to occupancy of all of the sixth floor of the bank structure. The minutes make no further mention of this move, but it must be assumed that it was carried out soon after approval of the lease on April 30, 1913.

Even the new quarters, however, could not quiet a growing dissension between several of the key staff members of the Exchange. For several months prior to February 18, 1912, trouble had been brewing between General Manager W. C. Temple and General Sales Manager R. P. Burton.

It was at the board meeting of February 18 that the matter was placed squarely before the board as a result of action by C. G. Harness, the cashier, in discharging an employee assigned to the sales department. Within the following few months, the increasing clashes between Temple and Burton had resulted in the resignation of both Temple and Harness, and the scars from this altercation were eventually in part responsible for the displacement of Burton as general sales manager some months later. The board eventually dissolved the office of general manager, and replaced that title with the title of secretary and business manager. Although completed action of the board took many months, the final alignment of key employees included E. D. Dow as traffic manager; and, later, F. L. Skelly, the general Northern manager, as general sales manager.

While this internal difficulty was festering in Tampa, progress was being made in the field of citrus canning. A report on record from F. Alexander McDermott of the University of Pittsburgh, working under the supervision of Professor Duncan, indicated progress in the preservation of orange juice in powder, syrup, and liquid form. This information, passed on to the board on April 2, 1914, moved that group to instruct McDermott to prepare, if possible, a sufficient quantity of preserved orange juice for the use of attendants at the convention of citrus growers which was scheduled for April 17 and 18, 1914.

Earlier, in March, the University of Pittsburgh had extended a fellowship to the Florida Citrus Exchange for its continued support of the citrus juice preservation program. The fellowship was in conjunction with the beneficence of A. W. and R. B. Mellon of Pittsburgh, who had established a permanent organization known as the Mellon Institute of Industrial Research of the University of Pittsburgh.

It is interesting to note that during this period the Exchange authorized the employment of William Wert of Pittsburgh in its Northern sales department. Wert, always a loyal and efficient worker, appeared on the November 5, 1912, payroll for the first time, and has continued this employment through the years. At this writing February, 1960, Wert is actively and efficiently conducting the office of division manager in the Exchange's Cincinnati office.

The annual meeting of June 3, 1913, saw few changes in the membership of the organization from the year prior. The reorganization of the board took place at a directors' meeting the next day. Dr. J. H. Ross was elevated to the presidency of the Exchange, and thus began his long executive service as another distinguished and highly respected president. W. B. Gray was elected first vice-president, and Dr. W. C. Richardson, second vice-president. Members of the board were G. W. Harp, H. R. Kenyon, G. W. Holmes, E. L. Pearce, W. T. Carter, L. W. Tilden, Edward Parkinson, George E. Koplin, J. W. Sample, W. E. Heathcote, S. C. Warner, H. L. Borland, J. P. Mace, Josiah Varn, and W. A. Fulton.

During the meeting William Hunter was renamed as legal counsel, L. D. Jones was reappointed secretary and business manager, and R. P. Burton continued for a short period as sales manager, with F. L. Skelly replacing him eventually, as indicated earlier.

The remainder of the fiscal year 1913 was marked by feverish action and several attacks on the Exchange system of requiring exclusive control of its members' fruit. To the credit of the board, it would appear from the minutes that each case was approached and ultimately settled without any relaxation of the contract specifications as established under the original charter and bylaws of the organization.

The annual meeting of the members of the Florida Citrus Exchange for 1914 was held on March 31. The membership roll at that time listed S. C. Warner, J. P. Mace, H. L. Borland, Ed Parkinson, L. W. Tilden, Dr. W. C. Richardson, Dr. J. H. Ross, G. W. Harp, G. W. Holmes, W. B. Gray, J. W. Sample, W. C. Temple, W. T. Carter, E. L. Pearce, H. R. Kenyon, W. E. Heathcote, Eugene Holtsinger, George E. Koplin, Josiah Varn, and W. A. Fulton. The Tampa employee payroll at the time included the names of R. P. Burton, L. D. Jones, W. T. Haizlip, E. D. Dow, W. T. Covode, H. T. Welch, H. C. Allan, A. E. Barnes, J. W. Carty, H. C. Plano, Walter D. Painter, Bruce Moseley, E. Burr, Mae Merzwick, Edward G.

Martin, Roy Oberholtzer, B. E. Merrill, Anna Pexa, A. Zeigler, P. P. Wood, Charles Metcalf, Blanche Gordy, C. V. Sylveria, William Gailliard, Bee Howk, Grace DeWolf, and D. H. Morton. This annual meeting took up and approved several articles of the charter and bylaws with amendments being presented by William Hunter, the attorney.

The board of directors' reorganization meeting took place on May 6, 1914; the board included Dr. J. H. Ross, president; W. B. Gray, first vice-president; Dr. W. C. Richardson, second vice-president; and L. W. Tilden, Edward Parkinson, J. W. Sample, H. L. Borland, W. T. Carter, J. P. Mace, E. L. Pearce, George Koplin, G. W. Harp, H. R. Kenyon, G. W. Holmes, S. C. Warner, W. A. Fulton, Josiah Varn, W. E. Heathcote, and Eugene Holtsinger. At this time there were twenty-three names on the Northern payroll of salaried employees, and an additional nineteen agents affiliated with the Exchange.

CHAPTER 4

1914-1915

EVENTS now moved quickly for the Florida Citrus Exchange, and the volume of business transacted by the board of directors reached such proportions that the minutes and other records were greatly abbreviated from what must have been the true coverage of matters of sufficient importance to come before the board.

Early in the 1913-14 season, the Thomas Advertising Service of Miami had been retained to represent the Exchange in all matters of advertising business, and the tone of the feeling among the board members seems to have been one of eagerness to advertise as much as possible in all markets where the Exchange either had or desired distribution.

It would appear that the board appointed prior to the beginning of the 1914 season saw fit to exercise more careful control over the Exchange administrative and sales staff than previous boards had chosen to do. Particularly in the expenditure of Exchange money, board action seemed to be required in almost every instance, with small margin left for the decisions of the business manager.

The 1913-14 season saw the Exchange more critically examining the grade of fruit sent out by its houses. There is evidence that some activity in this regard had been exercised the previous year, but in the 1913-14 season the Exchange demanded better grading policies from its associations. In addition, a memorandum of June 26, 1914, specified that all fruit shipped through the Exchange would be packed in boxes with the Exchange brand on the side, and that shipping of fruit in plain boxes would cease.

Also during this period the board decided to invite attendance of sub-exchange managers at all board meetings, a practice that was carried out for the next two years. These managers, in 1914, included C. H. Walker of the Polk Sub-Exchange; H. G. Gumprecht of the Manatee Sub-Exchange; A. B. Johnson of the Orange Sub-Exchange; H. N. Barnes of the Lee Sub-Exchange; W. W. Bateman of the DeSoto Sub-Exchange; and G. A. Neal of the Pinellas Sub-Exchange.

There is sufficient indication in the minutes of this era to suggest that the Exchange was beginning to experience difficulties in the handling of its associations through sub-exchange offices. There was also constant agitation regarding the methods employed by both the sub-exchanges and the associations in disposing of their members' fruit.

H. G. Gumprecht of the Manatee Sub-Exchange, for example, spoke for the director-representative of that group on January 6, 1915. He said that the Manatee board of directors was not satisfied with the method of distributing fruit. They felt, according to Gumprecht, that too much fruit was shipped to auction points, and some other system of marketing should be devised. They suggested that fruit should be sent to small towns and disposed of to individual dealers, and that a demand for fruit should be created in such towns by special advertising in the local papers.

On January 15, 1915, the board of directors decided not to attempt to handle sales of Florida's potato crop after a special committee appointed for this purpose reported that "for potent reasons" such an undertaking would be unadvisable. Later, on February 3, 1915, the board heard a letter from the Isle of Pines (Cuba) Growers Exchange asking that the Exchange send a representative to that area to investigate the possibility of the Florida Citrus Exchange accepting that group as a sub-exchange and undertake the task of selling their fruit. This matter was referred to the executive committee for further consideration, but the minutes of this period do not reflect any further action in this regard.

On April 14, 1915, a letter from the Boston district sales manager, George A. Scott, suggested that the implementation of a trade name for all Exchange fruit would considerably aid in the promotion of Florida citrus. Whereupon Jefferson Thomas of the Thomas advertising firm said that this matter had been under consideration, and that after many tests the trade name SEALD SWEET seemed to be the most suitable. It was promptly adopted by the board, and the proper officials were authorized to set about copyrighting it. The name itself was most outstanding, and it was to become a familiar trade-mark throughout the world. That it was durable is indicated by the fact that it remains today the master brand of the Florida Citrus Exchange.

Another historical event occurred at this same meeting on April 14, 1915. As a result of too many shipments of immature fruit both inside and outside of the Exchange system, various board members had for some time been expressing concern over this matter. At the April 14 meeting, the board finally agreed on a resolution directed to the Bureau of Chemistry of the Department of Agriculture at Washington requesting that the Bureau "fix a chemical standard of maturity which shall apply to inter-state shipments of all citrus fruits."

This action was apparently the forerunner of Florida's ever-tightening restrictions on maturity, and must certainly have met with much industry opposition at a time when restrictions of any sort were vigorously opposed.

The annual meeting in 1915 was held on June 1. Directors named at this meeting were L. W. Tilden of Orange County; H. L. Borland of Marion County; H. G. Putnam of Indian River Sub-Exchange; H. C. Hatton of DeSoto County; Dr. J. H. Ross of Polk County; Ed Parkinson of Lee County; J. W. Ponder of Manatee County; G. W. Holmes of the Highland Sub-Exchange; Dr. W. C. Richardson of the Hillsborough Sub-Exchange; Austin Roden of the Pinellas Sub-Exchange; and C. E. Stewart of the Volusia Sub-Exchange.

At an organization meeting, the newly elected board unanimously re-elected Dr. J. H. Ross to the presidency. Other officers included L. W. Tilden, first vice-president; H. L. Borland, second vice-president; L. D. Jones was elected to serve again as secretary and business manager and F. L. Skelly as general sales manager. William Hunter was reappointed as the attorney.

Tampa office employees at this time, in addition to the executive staff named in the foregoing paragraph, included W. T. Covode, H. T. Welch, H. C. Allan, B. E. Merrill, D. H. Morton, P. P. Wood, J. W. Carty, Grace DeWolf, C. V. Sylveria, W. D. Painter, A. E.

Barnes, E. G. Martin, H. C. Holmes, Frank Smith, J. R. Curry, C. H. Bucher, F. L. Hunter, Belle Sisk, G. H. Duggins, A. R. Sandlin, and C. C. Worthington. Northern sales office employees and agents were D. P. Kennedy at Atlanta, E. Temple Ware at Baltimore, George A. Scott at Boston, C. T. Allen at Boston, J. A. O'Malley at Buffalo, S. B. Wills at Chicago, William Wert at Cincinnati, J. B. Miller at Cleveland, J. Craig Allen at Dallas, W. H. Moody at Harrisburg, C. W. Chewning at Indianapolis, L. T. McGaughran at Memphis; R. H. Holland, J. Cohen, A. J. Casey, Joseph Connolly, all at New York, J. J. Reid at Peoria, C. A. Price at Pittsburgh, A. J. Neirmann at St. Louis, and C. N. Williams at Washington.

On November 17, 1915, recently appointed General Sales Manager F. L. Skelly requested permission to make the appointment of an Exchange representative in Detroit. The man he selected was H. C. McClaren, who remained as the district manager in Detroit until his death in 1951. He was replaced by his son, H. C. McClaren, Jr., who continues on the job held by his father.

During 1915 the question of shipments of immature fruit to the Northern markets plagued the Florida Citrus Exchange. A frequent topic of conversation at board meetings, the situation moved President Ross at one point to caution Exchange members against the shipment of green fruit. He then succeeded in ramming through a resolution that the Florida Citrus Exchange decline to handle any fruit that did not measure up to the standard specified, and that inspection be made by a person sent to the location by the Tampa office.

An item in the minutes of the October 20, 1915, meeting of the board, shows the breakdown of advertising expenditures by the Florida Citrus Exchange in concise and specific form. Jefferson Thomas, at this meeting, outlined his recommendations for expenditures during the rest of the year. They were adopted and approved as follows:

Wholesale and Retail Store and Price Cards Recipe Books, and Poster Stamps	$ 8,825.00
Magazine Advertising: *Cosmopolitan* *Good Housekeeping* *Literary Digest*	9,160.00
Streetcar Advertising	500.00

Additional Magazine Advertising: $1,079.00
Housewives League Magazine
Modern Hospital
Rotarian

General advertising expenditures, according to the minutes of this period, were apparently about $50,000 annually, and the assessment against sales of the sub-exchanges was firmly established at 16 cents per box, with all surpluses not exceeding 1 cent per box going into the emergency fund.

By April 16, 1916, the high prices paid by associations for supplies had caused discussion with regard to the relative necessity for forming some sort of central purchasing system which would place supplies in the hands of Exchange members at slightly more than wholesale cost. To this end, Dr. Ross, Dr. Richardson, L. D. Jones, F. L. Skelly, and W. B. Gray were appointed as a committee to investigate this possibility. In the following month the Exchange Supply Company was legally chartered, and the nucleus for the present-day Exchange Supply and Service Cooperative was founded in this action by the board of directors.

On June 8, 1916, L. D. Jones, the third top administrative executive of the Exchange, offered his resignation to the board. It was accepted by the board with expressions of regret. Jones apparently had served the board well, and he resigned with good feeling on all sides. Eventually named to fill the vacancy left by Jones was C. E. Stewart, Jr., of DeLand, manager of the Volusia Sub-Exchange.

This meeting, the annual reorganization meeting of the board, saw Dr. Ross re-elected to the presidency, J. W. Ponder chosen as first vice-president, and Z. C. Chambliss elected second vice-president. Staff officers remained the same as for the previous season.

As the Exchange prepared to go into the 1916-17 season, its board of directors consisted of Dr. Ross from Polk County, Z. C. Chambliss of Marion County, G. W. Harp of DeSoto County, L. W. Tilden of Orange County, J. P. Felt of Highland Sub-Exchange, J. W. Ponder of Manatee County, Dr. W. W. Birchfield of Pinellas County, Dr. W. C. Richardson of Hillsborough County, H. G. Putnam of Indian River Sub-Exchange, C. E. Stewart of Volusia County, and F. W. Perry of Lee County.

CHAPTER 5

1916-1917

THE Florida Citrus Exchange entered the 1916-17 season in what was considered by the board to be a sound financial condition. The organization was functioning smoothly and a general feeling of optimism is apparent from the recorded minutes of that era. In retrospect, the Florida Citrus Exchange was facing one of its most difficult seasons, but the board was of course unaware of this as it held its first meeting of 1916 on June 21.

It was at this meeting that the Lakeland Chamber of Commerce requested by letter that the Florida Citrus Exchange move its headquarters from Tampa to Lakeland. The board was disposed to invite the Lakeland group to appear at its next meeting for the purpose of further discussing this matter. Other community delegations were also invited to discuss the movement of Exchange headquarters to their particular localities. The matter of removing the location of headquarters to some other locality was put to a vote later in the year, and the motion was lost and considered dead at that time.

During this period, there is indication that the Exchange seriously considered the first of what would prove to be a long list of premium gadgets aimed at inducing the American housewife to purchase more citrus fruit. It is interesting to note that the Exchange board, at its meeting on September 20, 1916, decided to consider a fruit juice extractor which could be manufactured at low cost. During the following meeting, held on October 25, 1916, the board went on record as being opposed to the idea because of the high cost of manufacture. As one board member put it, "The Exchange is not a general merchandising company and I recommend that further consideration of the extractor be dispensed with."

The board declined during the October meeting to venture into the billboard advertising field. It did, however, make several motions intended to safeguard price information that was being distributed to sub-exchanges in bulletin form. As this information was somehow being "leaked" to competing sales agencies—to the great concern of the Exchange board—a special committee was appointed to look into this matter as well as to examine the entire scope of leakage of Exchange confidences.

On November 14, 1916, the Exchange received an invitation from the South Florida Fair and Gasparilla Carnival Association to provide a display and to furnish a citrus float. The board agreed to an expenditure of no more than $75 for this purpose, and thus established a precedent that was to exist for many, many years.

About a month later, on December 20, 1916, the board recorded the employment of H. G. Gumprecht, Jr., son of the Manatee Sub-Exchange manager, as assistant to the Chicago district manager. Gumprecht had gone to the Chicago market in August in order to acquaint himself with the complexities of auction selling. Forty-three years later, at the time of this writing, H. G. Gumprecht, Jr., is on the active list of key employees of the Florida Citrus Exchange, having served the organization continuously from 1916 to 1959.

The beginning of the new year found the Florida Citrus Exchange involved in several internal but routine difficulties, including certain problems in the bulky channels of communications that required all transactions with associations to pass through the sub-exchange offices. Most of these problems became unimportant, however, when Florida experienced the historic freeze of 1917 during the first week in February. Like all those in the citrus industry at the time, the Florida Citrus Exchange suffered greatly and was placed in the position of immediate retrenching of all activities.

A motion by F. W. Perry, seconded and unanimously adopted by the board at its first meeting after the freeze, is recorded in the minutes of the February 7, 1917, meeting. The motion outlined the confidence of the board in its executive staff in the following words: "The directors of the Florida Citrus Exchange hereby express their confidence in the executives of the organization and their approval of the handling of matters by their officials since the recent cold wave, and request that the board be furnished at its regular February meeting with an explicit statement of our financial condition and prospects and a recommendation that from this time on the management exercise such economy and retrenchment as seems feasible without crippling our efficiency."

No sooner had the industry adjusted itself to hard times incident to the freeze, which apparently reduced production to unimportant proportions, than another situation developed to further complicate both business and social progress across the country. For some three years the countries of Europe and Asia had been enmeshed in a great war triggered by conflict between Austria and Serbia. Slowly the conflict grew until it involved all the major nations of the world with the exception of the United States. Then, as the Germans began unrestricted submarine warfare early in February, 1917, the United States broke off diplomatic relations with the German government, which had declared war against France and Belgium. Great Britain had declared war on the Germans in 1914, and the Allies had waged the world's most enveloping war on record. But, with the beginning of Germany's unrestricted submarine onslaught, warlike incidents against the United States led to a declaration of war on Germany by the United States on April 6, 1917. These were times of great patriotism by Americans, and the Florida Citrus Exchange initiated a series of programs geared to assist the nation in its so-called "war to end war."

At the board meeting of April 25, 1917, the Exchange adopted and forwarded to the President of the United States the following resolution:

His Excellency, The President of the United States
Washington, D. C.

Your Excellency:

The Board of Directors of the Florida Citrus Exchange by resolution unanimously adopted today have the honor to tender the services of its organization in Florida and the principal cities of the United States, where we have salaried representatives, to be used by the Department of Agriculture, or other branches of the Government in the distribution of food products.

The Florida Citrus Exchange is a non-profit, co-operative organization of growers, organized for the purpose of marketing citrus fruits, the working force of which is composed of trained men of known efficiency in the distribution of foods. We have every reason to believe that their services would be of special value to the Government in this direction at this time.

If desired a representative will call on the proper authority to go into further details.

BY THE BOARD OF DIRECTORS
J. H. Ross, President

Following adoption of the above resolution at the April 25, 1917, meeting, the board also heard a short message by C. C. Hare, United States Department of Agriculture, with regard to the citrus industry's responsibility in time of war.

Following Hare's address, the motion was made, seconded, and passed that the board of directors of the Florida Citrus Exchange pass a resolution stating that it was the patriotic duty of every citizen of Florida to increase his efforts to produce more staple food products, that the board urge the Florida legislature to take some action looking to the creation of a commission to encourage the production and conservation of food products for the state of Florida, and that copies of this resolution be telegraphed to the Governor of Florida, the Senate, and the Speaker of the House.

It can be well understood that the Florida Citrus Exchange, following a disastrous freeze in February and this nation's entry into the war in April, 1917, was concerned about many phases of its operation. Because of this concern, a committee of board members, including J. W. Ponder, S. C. Warner, L. W. Tilden, G. W. Harp, T. L. Sams, and Dr. J. H. Ross, was asked to investigate the whole economy of the Exchange and to offer recommendations designed to see the organization through trying times.

On May 11, J. W. Ponder reported on the committee's findings:

Your committee spent the entire day considering the matters turned over to it for investigation. The present status of our organization was carefully gone into and the future needs and operations, based on the best information obtainable at this time, were carefully figured out from a most conservative standpoint.

It is impossible, at this time, to estimate with much certainty the amount of fruit that the Exchange will ship next season, but after considerable discussion of the information available, an estimate of 500,000 boxes was thought conservative. The following recommendations are made with this condition in mind.

The financial condition of our organization was carefully investigated, and although the exact amount of funds needed cannot be determined at this time, owing to the uncertainty of the amount of our shipments, still it is apparent that retrenchments must be made and the strictest economy practiced. Even with the following recommendations carried out, it will be necessary to borrow some money next fall to carry on the business until returns begin to come in in such amounts that will be sufficient to meet our expenses.

The committee wish to state that the work and services of all of our employees was fully considered and the service generally was very satisfactory.

Ponder then went on to recommend that the services of certain employees be dispensed with in view of the financial situation:

With the small amount of fruit there will in all probability be to ship next season, many sub-exchanges cannot be self-supporting and a combination of various sub-exchanges is recommended.

Much thought and due consideration was given to this matter and an arrangement was sought that would be of the greatest profit to all concerned.

It is recommended that Manatee, Hillsborough and Pinellas sub-exchanges be combined and that H. G. Gumprecht, present manager of Manatee Sub-Exchange, be placed in charge, and that Highland and Marion sub-exchanges be combined and that Mr. Barnes be placed in charge. This move will put efficient men in charge of these territories and sufficient fruit will come through these offices to meet the expenses this year. Of course, each sub-exchange would still maintain its organization and have its representation on the board.

The report of the Ponder committee was accepted by the board, and the business manager, C. E. Stewart, was directed to effect the recommended changes.

Thus the Florida Citrus Exchange concluded the 1916-17 season under difficult circumstances imposed by both nature and society. On June 5, 1917, the board for 1917-1918 was seated with Dr. J. H. Ross of Polk County, G. M. Wakelin of the Highland Citrus Sub-Exchange, D. C. Gillett of the Hillsborough Sub-Exchange, J. W. Ponder of Manatee County, Dr. J. E. Klock of Marion County, T. L. Sams of Indian River Citrus Sub-Exchange, J. A. Scarlett of Volusia County, L. W. Tilden of Orange County, and L. M. Hammel of DeSoto County.

Dr. Ross was renamed to the presidency of the board, L. W. Tilden was elected as first vice-president, D. C. Gillett as second vice-president, C. E. Stewart as secretary and business manager, and W. T. Covode as cashier. William Hunter continued as legal counsel, and F. L. Skelly was again named general sales manager.

CHAPTER 6

1917-1918

IN SPITE of the freeze and the war, the Florida Citrus Exchange seemed to bounce back into a relatively sound position by the close of the 1917-18 season. Minutes of the board meeting of June 20, 1917, show that there was considerable concern over whether $25,000 would be sufficient funds to properly advertise SEALD SWEET fruit.

Jefferson Thomas of the Thomas Advertising Service, who had directed the agency phase of the Exchange's advertising program since 1913, spoke at length to the board and there is indication that the board eventually decided that the $25,000 expenditure would be sufficient for the time.

Board member L. W. Tilden, at the same meeting, introduced the matter of Exchange sales of vegetables, long an issue of heated discussion among board members. From the minutes of this era, it seems apparent that several sub-exchanges had been handling vegetables for their citrus members. The question of handling vegetables obtained from all sources had been posed for many months, however, and Tilden asked for an organization of as many vegetable associations as possible to work under the same system used for fruit sales.

While Tilden's motion was seconded and carried, it is interesting to note that the September 26, 1917, meeting of the board passed a resolution requiring that purely vegetable growers must become members of the Exchange or of an Exchange organization before sales could be conducted for them. Thus, a member of an exchange association or sub-exchange for citrus could utilize Exchange facilities for the sale of his vegetables, but a vegetable grower not already affiliated with the Exchange would be required to become affiliated before the organization would undertake to market his vegetables.

On July 18, 1917, the Florida Citrus Exchange voted to join in the convention of the International Apple Shippers, and contracted for a half-page advertisement insert in the program of that organization. This initial participation in the apple convention was the forerunner of present-day participation which reaches considerable proportions. Through the years the Florida citrus industry has found it expedient to take part in the annual International Apple Association convention which is held during the short off-season of the citrus industry. The convention is an excellent opportunity for citrus interests to reach a great portion of the fruit trade and, most important, of produce buyers from all over the nation.

At the beginning of the 1917-18 season, the Tampa office payroll included C. E. Stewart, business manager; F. L. Skelly, sales manager; C. N. Williams, assistant sales manager; Edna Nall, stenographer; Frank Smith, file clerk; E. D. Dow, traffic manager; Grace DeWolf, stenographer, William T. Covode, cashier; W. F. Painter, mail clerk; J. Reed Curry, organizer; A. E. Barnes, inspector; J. B. Rust, inspector, H. C. Allan, auditor; and Jennie Smith, stenographer. William Hunter continued to act as the Exchange's legal counsel.

At its meeting on September 26, 1917, the Exchange issued the following resolution praising the work of the Florida State Plant Board: "Resolved that the directors of the Florida Citrus Exchange, assembled at their meeting this 26th day of September, 1917, herewith express their appreciation of the services of the Florida Plant Board and particularly of its director, Professor Wilmon Newell, in its energy and efficiency in its sphere of effort and notably in its splendid work in the control and eradication of citrus canker."

At this same meeting, a proposition was read to the board by the advertising committee recommending a special fresh fruit campaign in the *Literary Digest* magazine. The advertising cost was placed at $2,585, and the board voted in favor of the motion as presented by board member D. C. Gillett of Hillsborough County.

It was also at this meeting of September 26, 1917, that the board took up the problem of forming an organization for the purpose of providing financial aid to growers. Business Manager C. E. Stewart was instructed to pursue this matter, and it seems apparent that this action was to provide the foundation for the eventual organization of the present-day Growers Loan and Guaranty Company, which annually lends more than $3,000,000 to growers and shippers to assist them in producing and shipping citrus.

The matter of financing, always of prime interest to the board, had, during the first several years following the organization of the Exchange, taken several directions designed to provide financial assistance to growers. One such plan was the maintenance of Exchange accounts at banks throughout the citrus-growing area, particularly where sub-exchanges were located, so that bankers would look more favorably on loans to members of the Exchange.

This system apparently had its disadvantages, however, for on November 21, 1917, Dr. J. E. Klock, board member from Marion County, made the following resolution:

WHEREAS, the Florida Citrus Exchange has from time to time adopted different policies with respect to the distribution of its funds among different banking institutions, and

WHEREAS, the spreading out of these funds among a number of banks does not create for the Florida Citrus Exchange a sufficient account to warrant any one institution to extend the proper and necessary lines of credit the Florida Citrus Exchange at times needs, and

WHEREAS, it has come to the notice of the Board of Directors of the Florida Citrus Exchange that this condition of affairs is operating against the best interests of the Florida Citrus Exchange, now therefore be it

RESOLVED that we, the Florida Citrus Exchange, do hereby designate and appoint the Citizens Bank and Trust Company of Tampa, Florida, the official depository of this institution.

BE IT FURTHER RESOLVED that the Business Manager be and is hereby instructed to deposit in said bank all funds received by the Florida Citrus Exchange and disburse same by check, and to keep in said bank all balances of this institution (it being understood that this resolution has nothing to do with such deposits as the Florida Citrus Exchange may now have or in the future create relating to certificates of deposit drawing interest).

BE IT FURTHER RESOLVED that this resolution shall be binding upon the Florida Citrus Exchange as long as agreeable to both bank and Exchange and may be terminated upon notice each to the other of ninety days should either desire to discontinue, provided, however that should such arrangement be terminated the Florida Citrus Exchange will pay prior to such withdrawal any indebtedness it may owe said bank.

BE IT FURTHER RESOLVED that the Business Manager be requested to furnish the Citizens Bank and Trust Company with a certified copy of this resolution.

This resolution was carried unanimously by the board, and thus created a revised banking policy that became immediately effective.

In the minutes of the board meeting of December 19, 1917, can be found the complete list of newspapers on the advertising schedule of the Florida Citrus Exchange:

CONNECTICUT	*Bridgeport Post-Telegram, Hartford Courant, Hartford Times, New Haven Journal-Courier, New Haven Register, New London Day, Norwich Record, Stamford Advocate,* and *Waterburg Republican*
DISTRICT OF COLUMBIA	*Washington Star*
MAINE	*Portland Express*
MARYLAND	*Cumberland Times*
MASSACHUSETTS	*Christian Science Monitor, Boston Transcript, Fall River Herald, Lowell Telegram, New Bedford Standard and Mercury, Springfield Republican and News, Springfield Union,* and *Worcester Gazette.*
NEW YORK	*Albany Times-Union, Jamestown Journal, Rochester Union and Advertiser, Schenectady Union Star,* and *Troy Times.*
PENNSYLVANIA	*Allantown Call, Chambersburg Repository, Harrisburg Telegraph, Lancaster Examiner, Lebanon Report, Reading Eagle, Shamokin News, Williamsport Sun,* and *York Dispatch.*
RHODE ISLAND	*Providence Journal and Bulletin*
WEST VIRGINIA	*Wheeling News*
CANADA	*Toronto Star*
COLORADO	*Denver Post*
ILLINOIS	*Bloomington Pantagraph, Danville Com-*

	mercial News, Peoria Journal, and *Springfield State Register.*
INDIANA	*Indianapolis News*
MICHIGAN	*Detroit News, Detroit Times,* and *Grand Rapids Herald*
MINNESOTA	*Minneapolis Tribune*
MISSOURI	*Kansas City Star*
NEBRASKA	*Omaha News*
OHIO	*Columbus Dispatch, Columbus State Journal, Dayton News, Springfield News, Toledo Blade,* and *Zanesville Times-Recorder.*
TEXAS	*Dallas Journal, Dallas Times-Herald*
WISCONSIN	*Milwaukee Evening Wisconsin*

These newspapers were scheduled to insert advertising in the amount of $18,861.23 during the remainder of the season from December 19, 1917.

With regard to advertising by the Florida Citrus Exchange at this time, General Sales Manager F. L. Skelly presented to the board on March 20, 1918, a long oration on the status of Exchange sales during the 1917-18 season. Skelly's address seems to have so thoroughly covered both sales and advertising, and to have been so comprehensive in nature that the entire speech is carried verbatim.

TO THE BOARD OF DIRECTORS OF THE FLORIDA CITRUS EXCHANGE:

I believe this the proper time to present to you a partial review of our work for this season, and also to try to point out some of the lessons it has taught us.

The sales department has had its troubles and has not been able to accomplish all of the things which we desired to; however, I am well satisfied with the record we have made this year. My satisfaction would be much greater if we could use what we have learned in making plans for the future that will produce even better results.

It is not necessary to go into details regarding the things that have helped most and hindered most, as the board of directors have been well informed about same as they came up and were disposed of. In order to make this paper as brief as possible, we will only touch on matters which really are worthy of our mention.

You all know a short crop is almost as much of a handicap to our work as it is an advantage. When fruit is scarce it is undoubtedly easier to secure good prices provided the right kind of distribution can be had. The question of getting this distribution is not to be lightly considered, as I have always contended and as we have had ample evidence this season. The quality of considerable of the fruit which it was necessary for us to market, the transportation tie-ups and extremely cold weather for two months have combined to keep us very busy. We also found that considerable fruit was more damaged by the cold than many of the growers thought. This fruit going in to the markets dry and unfit for consumption caused us considerable trouble and correspondence; however, we got almost all of the complaints adjusted satisfactorily. The balance of them, as you know, we turned over to the auditing committee at yesterday's meeting.

Taking everything into consideration I do not believe there can be any complaint regarding prices obtained for oranges that were handled properly by our associations in this state. We had some low sales on oranges, it is true, but these were either frozen oranges, fruit improperly handled at the packing house or transferred cars which no doubt caused much of the decay.

Good grapefruit has sold at satisfactory prices this season. We had some grapefruit during the extremely cold weather that did not bring the prices we would like to have had; however, taking the sugar shortage, transportation and cold weather into consideration, I feel the results on grapefruit should be considered satisfactory, although grapefruit did not rule as high correspondingly as oranges. This may be traced to the extremely cold weather and, as stated above, the sugar shortage; also to the fact that there was a great deal more grapefruit in Florida than early estimates indicated and to the fact that grapefruit is not nearly as well known to the consuming public as oranges.

Because most people are pretty well acquainted with oranges, our educational campaign on the food and health values of citrus fruits has been applied by the general public much more to oranges than grapefruit. Since the California Fruit Growers Exchange started advertising to emphasize these same points, following our lead, they have been even more generally understood to be true only of oranges, as our California friends advertise nothing else.

The result has been that the average housewife has come to look upon oranges as well worth using from the food and health standpoint. This year we were able to cash in on this feeling to an unusual extent because of the shortage of California fruit during our season. If we are to continue to profit by it, however, we must strongly advertise the superior merits of Florida oranges. Otherwise with a normal volume of California fruit and the extensive advertising which the California Exchange does, the public will largely forget that there are Florida oranges in the market.

A careful study of the situation convinces me that a great deal of the fruit we have sold has gone to the homes of laboring people who are now receiving high wages. On the other hand, the homes of people of means, in which oranges have heretofore been freely used, probably have not taken as many as in the past. In other words, I agree with Mr. Hoover that there is a great deal more food conservation practiced in the better class of homes than in those of the people whose buying capacity in the past has been limited.

In almost every home in the United States, some oranges have been used in the past, at least on special occasions. The members of the family know something about the fruit and when you talk about oranges they can understand you. Consequently, the advertising of oranges as food and for health purposes which has been done both by the Florida Citrus Exchange and the California Fruit Growers Exchange was read by the people in receptive moods. Many of them felt in a way they were contributing to the conservation of staple foods by using more oranges, just as we have been teaching.

With grapefruit, conditions have been different. We must face the fact that comparatively few people really know what grapefruit are. Of course hundreds of thousands of persons are familiar with this fruit, but there are millions who are not. The great bulk of the past sales of grapefruit have been in families in fair circumstances. They have been looked upon as a luxury, and when war economies began to be practiced, a great many people in good circumstances dropped grapefruit for this reason.

Because so few people really know how good grapefruit is, or that they have much the same food and health qualities as oranges, there was not the increased consumption of them in the homes of highly paid laborers that there might have been. We must not forget that grapefruit has been pushed only by the Florida Citrus Exchange previous to this season—when the Puerto Rican growers put on an advertising campaign for this fruit, whereas oranges have had the benefit of our publicity and also of the much more extensive campaigns of the California Exchange.

The main features of our work this year, as I see them, are taking advantage of existing conditions to introduce Florida oranges into communities heretofore dominated by California fruits, the increased proportion of our shipments which have gone in private sale markets as compared with auctions, and the inauguration of the contract basis for orders with advance deposits required.

Frankly, in contending that the latter is the greatest advance step ever taken in the merchandising of citrus fruits, I am not expressing my own opinion but giving the practically unanimous judgment of men familiar with selling problems who have informed themselves regarding this matter. I am hopeful that the benefits derived from this method of selling fruit will be so clear to our growers that they will feel justified in letting us go much further in adopting the practice of modern selling methods.

As already indicated, I believe that to hold the place we now have in the sale of oranges and to increase our volume of business to correspond with present prospects for next year's crop, we must push oranges even harder than we have been. By keeping before the public the superior merits of Florida oranges, we will be able to hold our own and greatly increase our sales in sections which have been almost exclusively California points.

There is no doubt that oranges will be eaten more and more freely by the American people every year, and it is up to us as to how far this increased consumption will benefit Exchange growers. With oranges, we can protect and push the interests of our growers through the methods heretofore followed, if these are prosecuted with sufficient vigor. I am sure we will have to advertise SEALD SWEET oranges more extensivley and cover the Northern territory with more men in order to get our share of the business.

I have no radical innovations to suggest as to our methods of advertising or the way in which our men shall work. If we have the money and men, there is no doubt that we can keep up our end with California, even though she has greater resources and greater volume of trade. The fact that we have the better goods surely will help us a lot, if we grade, pack, etc., in the right way.

As regards grapefruit, we have a different and more difficult problem. As I have tried to point out, the American public, as a whole, must be educated as to what grapefruit really are and we have to show the dealers that this fruit can be sold in large quantities if properly pushed. In a large section of the country our work must be opposed to that of the Puerto Rican growers who are advertising freely and in many cases consigning their fruit.

For our grapefruit campaign therefore, I would recommend and urge a more aggressive policy. If present crop prospects are realized there will probably be 4,000,000 boxes of grapefruit in Florida next fall. The chances are we shall be called upon to market at least 800,000 to 1,000,000 boxes. Unless we take some positive steps to further popularize grapefruit and standardize prices for it, I am afraid many growers will be disappointed in their returns.

And if next year's crop is four million boxes, barring freezes and disasters of like nature, it is not likely to be many years until we are confronted with perhaps a crop two or three times as large.

It seems to me that we have to get right down to brass tacks and put on a real campaign for grapefruit, just the kind of campaign that the manufacturers of a new substitute for lard or something of that kind would certainly undertake. I believe that we have got to introduce grapefruit in the nine out of ten homes where it is now scarcely known. To do this we must advertise liberally to the housewife, and put vigorous merchandising methods in our work with the trade. I have pretty well defined plans in my mind as to just how this shall be done, but I do not want to worry you with the details at the present time. If you will endorse the general policy, take such steps as will insure the necessary money, and authorize the sales and advertising departments to work out these plans fully during the next few weeks, I am sure that when finally presented to you they will meet your hearty approval.

To make the thing very concrete for your consideration, I am going to ask you right now to begin to plan for $100,000 with which to push next year's grapefruit crop, if our part of it is a million boxes. It may seem to some of you that I am talking in big figures, but I think I know what we have to face and what we need to make our fight with.

I don't mean to recommend that this entire amount be spent for advertising—after going into the matter very thoroughly with our advertising agency, I believe the plans we would mutually agree upon would probably put not more than half of the amount into newspaper space.

If we undertake the job of selling a million boxes out of four million boxes of grapefruit produced in Florida without the advertising and merchandising I am asking you to authorize, the average price per box will, in my judgment, be from twenty-five to fifty cents lower than can be obtained if we put on the proposed campaign. If we do not get busy and open up new avenues of consumption for grapefruit, while the buying capacity of the people of the United States is at a high level, with thousands and thousands of acres of groves that are coming into bearing this fruit, in a few years will not pay the grower anything for his time, labor, and investment.

Is it not worth while to spend at least ten cents a box to increase next year's returns by two to five times this sum and build up such a demand for grapefruit as will afford an avenue through which can be profitably sold the production of the State even when this has become many times as large as it will be this year?

In my opinion no time should be lost in getting busy on some such lines as I have tried to suggest in this paper. We cannot afford to delay in engaging men. The extraordinary conditions now prevailing make them scarce and hard to get.

Some of our best district managers already have been approached with liberal propositions to take other positions. If we are to maintain our sales force, in this office and in the North, if we are to expand it and make it more efficient, we must face conditions as they exist and meet them as other enterprises are doing.

Should it be decided to put on an aggressive educational and sales campaign for grapefruit, the success of this will to a very large degree depend upon starting early and working out our plans carefully and with due consideration of all factors in the situation.

You have had experience with the amount of time required to get up good advertising matter, and under existing conditions it will take longer than ever before. Money can be saved and the work of every man we employ made more effective if we start early and take advantage of all the ins and outs of the merchandising and publicity end of our endeavor.

I am confident that the further we go in this direction the easier it will be to get more fruit into the Exchange and to hold our old growers.

I trust you will see your way clear to give us authority to go ahead and work out plans that we may obtain results the coming season that will be satisfactory to our growers and yourselves.

Although Skelly's suggestions were not immediately accepted, it will be seen in later chapters that his ideas penetrated into the thinking of the organization and became eventually quite useful.

CHAPTER 7

1918-1921

AS COULD be expected, the continuing importance and evident success of the Florida Citrus Exchange presented inherent problems to the organization. One such problem was the tendency of the other sales agencies to imitate closely or copy the SEALD SWEET trade name. Another problem was the protection of the reputation of the Exchange for quality. In order to lay down policy within the Exchange organization in this matter, President J. H. Ross made the following observation on April 17, 1918:

The resolution asking how we may safeguard and protect the reputation of our trade name SEALD SWEET is timely. The word was selected for the impartial benefit of all growers affiliated with the Florida Citrus Exchange. Its ownership is vested in the Florida Citrus Exchange which consists of the directors chosen by the subexchanges. The Florida Citrus Exchange is a marketing agency.

In performing the function for which it exists it must do all the things necessary to the highest efficiency in selling the products of the affiliated growers.

Among the things inseparable from selling in the most efficient manner are maintaining a marketing force here and throughout the country involving maintaining offices, clerks, stenographers, bookkeepers, inspectors, draying facilities, diverting points and agents, advertising, etc.

The motion calling for this report touches the matter of advertising and efficient selling directly. Since the Exchange had adopted the policy of advertising, it became logically necessary in order that the advertising should benefit particularly the affiliated growers that a trade name be adopted to designate their fruit in the markets to distinguish it from fruit not affiliated with the Exchange.

So the trade name SEALD SWEET was adopted. In the effort to popularize SEALD SWEET fruit we have paid thousands of dollars in describing the kind of quality of fruit we profess to offer the public under the SEALD SWEET designation. All this was wise, businesslike and along the lines of scientific marketing.

It would seem from all this that the Florida Citrus Exchange, owning this trade name, using it for purposes of description of fruit offered for sale, guaranteeing the quality under this name in all our advertising, must logically, inevitably, and undoubtedly have the authority to say what quality of fruit may be shipped and offered for sale as SEALD SWEET.

We assume that it has this authority, as it owns the name and is charged with the duty of marketing the fruit of its affiliated associations to the best advantage and as to the details of how this may best be done, it is and must be the sole judge.

No grower, no association, not even a sub-exchange can dictate these details of selling methods or the methods themselves.

When they are dissatisfied they may be heard through their representatives in the Exchange, and if the Exchange cannot satisfy the complaining of dissatisfied grower, association or sub-exchange, such grower, association or sub-exchange has the right to withdraw his or its affiliation with the Exchange.

One grove, one portion of a county, or a whole county or a larger region may suffer seasonable damage by a hot wave or a cold wave which results in greatly inferior fruit, fruit far below the high standard fixed by advertising as SEALD SWEET fruit.

In such circumstances, indicated by general knowledge of temperatures or other adverse influences, the Florida Citrus Exchange must repose authority in its officers to employ inspectors to report upon such fruit suspected of being deteriorated or maybe damaged, and if such a report by inspectors be adverse, verified by other inspectors if the grower or owner of the fruit so desires, the Florida Citrus Exchange may and should decline to permit such damaged or deteriorated fruit to carry the trade name SEALD SWEET.

On June 6, 1918, the board met in its regular annual meeting and the credentials of board members were accepted. The board included Dr. J. H. Ross of Polk County, L. M. Hammel from DeSoto County Citrus Sub-Exchange, G. M. Wakelin of Highland Citrus

Sub-Exchange, D. C. Gillett of Hillsborough County Citrus Sub-Exchange, H. G. Putnam from Indian River Citrus Sub-Exchange, D. S. Borland of Lee Citrus Sub-Exchange, J. W. Ponder of Manatee County Citrus Sub-Exchange, J. E. Klock of Marion County Citrus Sub-Exchange, P. C. Peters of Orange County Citrus Sub-Exchange, and A. G. Hamlin of Volusia County Citrus Sub-Exchange.

With regard to the election of P. C. Peters to the board of directors, it is interesting to note that as of February, 1960, Peters was serving as the president of the board of the Florida Citrus Exchange. The intervening years between his election to the board in 1918 and the present have provided him an unchallenged position at the head of the Exchange and as an industry leader.

The board was reorganized with the re-election of Dr. Ross as president, D. C. Gillett as first vice-president, L. M. Hammel as second vice-president, C. E. Stewart, Jr. as secretary and business manager, Judge William Hunter as attorney, and F. L. Skelly as general sales manager. E. D. Dow was elected as traffic manager and W. T. Covode as cashier.

The Tampa payroll on July 17, 1918, included, in addition to the salaried officers listed above, Jennie Smith, C. N. Williams, Edna N. Morrill, Mrs. H. E. Kerr, Frank Smith, Blanch Dossel, Grace De Wolf, Julia Grace, B. C. Frazer, H. F. Phillips, Mrs. Almira Midgley, H. C. Allan, W. D. Painter, J. Reed Curry, A. E. Barnes, J. B. Rust, C. A. Price, H. E. Wescott, W. L. Harper, Louise C. Flisch, and Y. Briddell.

While the optimistic beginning of the 1917-18 season had given way to much concern over the future of the Exchange because of the war and disastrous weather, the rapid comeback of the organization is apparent in an item from the July 17, 1918, meeting of the board. Thc salary committee of the board in an unprecedented move at that time recommended an increase of 10 per cent in the salary of every employee of the Florida Citrus Exchange on the Tampa office payroll. About two months later, on September 18, 1918, the board, faced with a serious shortage of pickers because of the war effort, voted to contact the proper authorities for soldier help in harvesting the coming crop. The appeal to Washington was not, as it turned out, to be pursued to its ultimate objective because of the rapid progress of the war. With the signing of the Armistice on November 11, 1918, the Florida citrus industry began the difficult conversion from war to peace.

One of the first conversion problems to be encountered was in the matter of transportation. The board deadlocked in the consideration of a resolution proposed by D. S. Borland which would urge the government to return all transportation lines to their owners at the earliest possible moment. The deadlock was eventually broken by Dr. Ross in favor of the resolution, and each member of Florida's congressional delegation was presented with a copy of the Exchange's policy regarding continued government control of transportation. The congressional delegation receiving copies of the Exchange resolution were Senators Duncan U. Fletcher and Park Trammell, and Representatives Herbert Drane, William Sears, Frank Clark, and Walter Kehoe.

A report of movement of fruit by the Exchange in the 1917-18 season up to December 15, 1918, is recorded in the minutes of the December 18, 1918, meeting of the board. The movement of citrus by the Exchange was 599,000 boxes, which compared rather favorably with the preceding season, during which the Exchange had moved only 252,059 boxes in the corresponding time.

During this period, an address by General Sales Manager F. L. Skelly seems to be indicative of the success of the Exchange's entire operation:

> You may have noticed that we continue to sell grapefruit and oranges for considerably more money than the prices offered by outside operators. It is to be assumed that the bulk of fruit in the hands of the "fly-by-night" speculators at the beginning of the season has been moved or will shortly go out. The better-established competitors of the Exchange seem to feel that they are doing well to obtain within fifty cents to a dollar a box of our prices. That this is true shows just how well it pays to market cooperatively fruit carefully packed and well advertised.
>
> It is gratifying to be able to say that our distribution of SEALD SWEET fruit up to now has been the widest in the history of the Exchange. In consequence of being able to get a greater volume of fruit into outside markets it has been possible to keep our shipments into the auction markets down; and we have put very little fruit into the various auction markets. Inasmuch as our circularizing and trade advertising is just under full headway, it seems reasonable to expect continued good results from this work.

Minutes of the several board meetings in the final months of 1918 indicate the board was actively engaged on several different projects including plans for legislative action to help in the fight against citrus canker, the solid formation of the Growers Loan and Guaranty Company for financial assistance to growers and shippers, and a move to require the Thomas Advertising Service, the Exchange's agency, to create a branch office in Tampa to handle the Exchange's business.

One project of the Exchange at this time was the precooling of citrus before shipping it to market. Various house experiments in 1917 and 1918 had already proved successful in the matter of extending the shelf life of fresh citrus. In March, 1919, S. J. Dennis and A. W. McKay, both of the Bureau of Markets, United States Department of Agriculture, addressed the board concerning the advantages of precooling. The reception of the address is indicated in the minutes of March 19, 1919, board meeting which reveal that ". . . the directors and sub-exchange managers showed a keen interest in what Mr. Dennis had to say and asked many questions, and the discussion was prolonged for some time."

The board for the 1919-20 season, which was seated on June 3, 1919, was composed of the following representatives of citrus sub-exchanges: J. H. Ross of Polk County, C. E. McCormick of the DeSoto County Citrus Sub-Exchange, A. G. Hamlin of the Volusia County Citrus Sub-Exchange, P. C. Peters of the Orange County Citrus Sub-Exchange, Edward Parkinson of the Lee County Citrus Sub-Exchange, H. G. Putnam of the Indian River Citrus Sub-Exchange, J. E. Klock of the Marion County Citrus Sub-Exchange, J. W. Ponder of the Manatee County Citrus Sub-Exchange, G. M. Wakelin of the Highland Citrus Sub-Exchange, and D. C. Gillett of the Hillsborough County Citrus Sub-Exchange. Officers and key executives for the coming season were also elected during this meeting. Dr. J. H. Ross was re-elected to the presidency, J. W. Ponder was elected as first vice-president, and D. C. Gillett was elected as second vice-president. C. E. Stewart, Jr. was renamed secretary and business manager of the Exchange, W. T. Covode was renamed as the Exchange cashier, E. D. Dow was reappointed as traffic manager, and William Hunter was reappointed as attorney for the Exchange.

F. L. Skelly, who had held the position of general sales manager for several years, was replaced by George A. Scott, the Exchange's Eastern division manager prior to his appointment as general sales manager. While neither the regular minutes nor the executive session minutes of the period reflect the reasons for Skelly's release as sales manager, citrus veterans who remember the era are agreed that Skelly asked that he be allowed to resign from his position in order to take a similar position with another organization. The board action on June 3, 1919, concluded his association with the Exchange.

On July 16, 1919, the board approved the advertising budget for the 1919-20 season as presented by the Thomas Advertising Service. The total budget was, of course, allocated to written media and amounted to $97,000 for the year. Nearly a third of that amount was allotted to magazines, including the *Literary Digest, Saturday Evening Post, Delineator, Good Housekeeping, Ladies' Home Journal, Pictorial Review, Today's Housewife,* and *Woman's Home Companion.* The balance of the advertising budget was earmarked for newspapers, special promotions, circulars, point-of-sale pieces, and advertising production costs.

On November 19, 1919, the board heard a recommendation from both the Sub-Exchange Managers Association and the Exchange Supply Company that the Florida Citrus Exchange establish its own laboratory for the purpose of conducting tests of fruit shipped under the SEALD SWEET brand name. The board indicated its agreement and named J. W. Ponder, Dr. J. E. Klock, H. G. Putnam, Edward Parkinson, and S. C. Warner as a committee to investigate and pursue the matter of establishing a laboratory.

Just before the close of the year on December 17, the board again heard an offer involving the relocation of Exchange headquarters. P. C. Peters told the board at its December 17 meeting that the city of Orlando had offered to give the Exchange a lot of suitable size for the construction of an office building in Orlando if the board decided to move its general offices from Tampa to Orlando. As the minutes note:

> . . . The above motion brought before the board the matter of the advisability of moving the Exchange headquarters and erecting a building that would house the Florida Citrus Exchange, the Exchange Supply Company, and the Growers Loan and Guaranty Company, and, upon motion seconded and duly carried the president was instructed to appoint a committee of five, from the board of directors, to investigate the advisability of moving the Exchange headquarters, and also the advisability of erecting a building.

The committee included H. G. Putnam, P. C. Peters, J. W. Ponder, D. C. Gillett, and Edward Parkinson. There is no indication that the matter was given much more than token consideration at this time.

It is interesting to note that the Prohibition amendment to the United States Constitution had been ratified at this time and was scheduled to go into effect on January 16, 1920. The Florida Citrus Exchange could see increased emphasis on fruit drinks in prospect as the nation began its switchover from alcoholic to nonalcoholic beverages. Business Manager C. E. Stewart, Jr., made reference to Prohibition in the text of a long report to the board submitted on December 17, 1919. The reference reads:

> Prohibition has caused an enormous interest in fruit drinks, and particularly those made of citrus fruits, and this fact alone bids fair to greatly increased demand; and the outlook, from all information that we could gather prior to shipping, indicates good prices.

Soon after the beginning of the year 1920, the Florida Citrus Exchange became interested in a school program designed to reach the coming generation of housewives with the virtues of including citrus in their menu planning. On March 3, 1920, the business manager brought before the board of directors the matter of the successful demonstration under the direction of Mrs. Caroline Moorhead in Tampa at the South Florida State Fair from February 16 to February 21 (1920) and requested from the board the privilege of making some experiments to prove the value of demonstrating the use of citrus fruits, particularly grapefruit, to the home economics classes in some of the Southern colleges. The board approved this plan and the business manager was instructed to proceed with such plans as he thought proper in carrying out such a program.

Thus the Florida Citrus Exchange entered the fabulous 1920's. In retrospect, the first eleven years of the Exchange had provided a foundation and scores of policy precedents that were to prove particularly beneficial during the ensuing ten-year period. All of this was in the future and known by no mortal man on June 1, 1920, when the 1920-21 board presented its credentials and was seated. The new board consisted of Dr. J. H. Ross of Polk County Citrus Sub-Exchange, J. W. Ponder of Manatee County Citrus Sub-Exchange, D. C. Gillett of Hillsborough County Citrus Sub-Exchange, H. G. Putnam of Indian River Citrus Sub-Exchange, A. G. Hamlin of Volusia County Citrus Sub-Exchange, P. C. Peters of Orange County Citrus Sub-Exchange, W. W. Raymond of Lee County Citrus Sub-Exchange, and C. E. McCormick of DeSoto County Citrus Sub-Exchange.

Re-elected to the presidency was Dr. J. H. Ross. D. C. Gillett was elected as first vice-president, J. W. Ponder was chosen as second vice-president, C. E. Stewart, Jr., was renamed as secretary and business manager, W. T. Covode was renamed as cashier, and William Hunter was reappointed as attorney for the Exchange.

CHAPTER 8

1921-1922

THE PERIOD from August, 1921, until August, 1922, was an active and extremely productive time for the Florida Citrus Exchange. Members of the board during this period were Dr. J. H. Ross of Polk County Citrus Sub-Exchange, A. G. Hamlin of Volusia County Citrus Sub-Exchange, J. W. Ponder of Manatee County Citrus Sub-Exchange, P. C. Peters of Orange County Citrus Sub-Exchange, D. E. Gillett of Hillsborough County Citrus Sub-Exchange, Dr. Y. E. Wright of DeSoto County Citrus Sub-Exchange, T. L. Hausman of Indian River Citrus Sub-Exchange, C. J. Stubbs of Lee County Citrus Sub-Exchange, Walter R. Lee of Marion County Citrus Sub-Exchange, and F. C. W. Kramer of the Highland Citrus Sub-Exchange. All staff officers remained unchanged at the beginning of the 1921-22 season.

The events and progress of the Exchange as contained in this chapter with regard to the 1921-22 season are taken from one of the few early, complete annual reports of the business manager. They have, of course, been documented by constant reference to the official minutes of this period, but the report covering this period is quoted in the words of Business Manager C. E. Stewart, Jr.:

Practically all farm products, during the past year, sold at unsatisfactory prices. The readjustment of general business created a condition which affected the sale of these products. Citrus fruit from Florida, however, was an exception. The first part of the season, it sold at satisfactory prices with every indication pointing to a reasonably good season, because although the country was being threatened with strikes, business was undergoing drastic liquidation and the process of reconstruction was in full swing, it was soon realized that the methods of this organization were being felt in the markets more than ever before. This was because our methods under our new sales manager were being changed gradually during the two years previous, and a more aggressive policy used in our selling.

The development of new markets last season was the greatest of any year in the history of the Exchange. The Exchange was the greatest factor in the markets, last year; it soon got control of the situation and worked to stabilize prices. Our wide distribution tended to diminish slumps. On January 19, 1922, a freeze in California practically stopped shipments from that state and so damaged that crop that they were not a material factor for the remainder of the season.

This condition was reflected immediately in prices realized on Florida citrus. The Exchange at this time had the greater portion of its holdings to move and members were able to realize much greater returns for their crop.

In a year when business was in the condition above mentioned the advantages of cooperative marketing, making for orderly distribution, were apparent. Undoubtedly the greatest factor in stabilizing the market before the California disaster, and the greatest factor in getting the high dollar after the California freeze, was the Exchange.

The Florida crop amounted to 33,023 cars, made up of 14,930 cars of oranges, 18,093 cars of grapefruit. The Florida Citrus Exchange handled 10,572 cars, or 32 per cent or one-fourth of 1 per cent less than the percentage controlled last year; this loss was occasioned, principally, by the storm of October 25, 1921, and the losses to growers were heaviest in those sections where we had made our best gains. Further, the storm affected grapefruit to a considerable extent in Polk County, causing losses which continued throughout the year.

Prior to this storm all sub-exchanges had reported very material gains, and we have reason to believe that these estimates were accurate. However, the final results show gains only in Dade, De Soto, Indian River, Lee, and Volusia, amounting to 367,053 boxes, and losses in Highland, Hillsborough, Manatee, Marion, Orange, Pinellas, and Polk, amounting to 466,952 boxes. Two main conditions affected the percentage of our crop.

First, a change of managers in the Highland Citrus Sub-Exchange and Marion County Citrus Sub-Exchange probably was the cause of a loss there of 190,930 boxes. Second, although Orange County reported a considerable gain in membership, they were confronted with a short crop of oranges in the very territory where they had made their gains.

Organization Department

Mr. J. Reed Curry is chief organizer, having held that position with the Exchange for seven years.

The many duties embraced in the work of the Organization Department appear to the observer only after he has become thoroughly familiar with the comprehensive plans and extensive efforts being made, through this department, to accomplish more complete cooperation among the growers of Florida.

Its operations are not only necessary in the progress of our business, but are, in reality, fundamental; and to a large degree the success of the Florida Citrus Exchange is dependent upon the initial work done by this department through the main office and the various sub-exchanges and associations. Additional volume of fruit and new members must be obtained each year throughout the state in order that the Florida Citrus Exchange may grow larger and increase its control of fruit production of the state. In accomplishing this, the growers must be visited and their affiliation obtained.

While it is true that much has been written and published regarding the methods and purposes of the Florida Citrus Exchange, and it would therefore seem reasonable that by this time all growers of the state would fully understand and appreciate the value and need of it, yet it is a fact that a very large percentage of the growers, both in the organization and out of it, have not yet comprehended the system of cooperative marketing in all its important details.

Therefore, one of the main objectives of the Organization Department is to meet such growers individually and collectively, and, by patient and exhaustive effort, explain what the Florida Citrus Exchange is, why it exists, how it is formed, how it is operated, methods necessary to increase demand and build up markets, show the needs of grove records and the production of high grades of fruit, and all other things which affect the welfare of the business.

It is readily seen that much of the work of this department is EDUCATIONAL, and in order to get best results close attention must be given to the differences in personality of the growers, their peculiarities, environments, and past experiences. In approaching them upon the subject of cooperative marketing, it is necessary that this department should have a complete knowledge of their problems, and be actuated by an honest desire to assist in solving them.

In visiting hundreds and hundreds of growers during the past year, we have found it advisable to spend hours with many of them in order to clear their minds of misunderstanding or prejudice, and convince them that their interests are dependent upon our success, and that their cooperation would be of advantage to themselves and to us.

One of the functions of this department is to get together the growers of new communities, organize them into associations, assist them in plans for financing and building new packing houses, and visit them at regular intervals thereafter, to help them to operate successfully.

Your organizer attends as many associational meetings as possible, to address the members upon subjects of interest and inform them regarding new methods, and also keep them advised regarding progress of other associations and the general work throughout the state. During the past year he has attended 73 meetings of members and directors of associations and sub-exchanges.

Sometimes it becomes necessary to adjust differences which have arisen among the members of associations, and this particular duty calls for the most diplomatic and careful efforts, since to take sides with one faction will invariably antagonize the other.

The effort therefore must always be to appeal to the loyalty of the members as a whole, and if possible effect a compromise of the difficulties.

During the past year it has seemed that the competitors of this organization have been unusually active in spreading reports and statements that were false and misleading. To offset and correct such propaganda the organization department has been compelled to follow up and trace down reports. It has always found them to be started by speculators, whose motives were to discredit the Florida Citrus Exchange.

New associations have been organized during the past year in such localities as needed them, and there are already prospects for several additional packing houses for next season.

In connection with the organization work of this department your organizer has travelled, during the year, approximately 28,000 miles, and has visited every important citrus section of the state.

Because a large percentage of growers are non-residents, it has been necessary, at times, for visits to be made to Northern states to secure their cooperation.

Such recommendations as might be made by this department would include more active efforts to reach the non-resident growers, and to acquaint them more fully with the work of this organization.

Taking the state as a whole, your Organization Department feels much gratified over the present interest and enthusiastic cooperation of the entire membership, and considers the conditions at this time to be more favorable than ever before.

TRAFFIC DEPARTMENT

The Traffic Department is under the direct management of Mr. E. D. Dow, who has held this position for 10 years. Mr. Dow has been in the employ of the Florida Citrus Exchange for 13 years and, by a few months, has the longest record of any employee.

During the shipping season of 1921-22, there was a noticeable improvement in the service rendered by the carriers throughout the country, in that shipments were handled more expeditiously. An adequate car supply was available throughout the season, and a noticeable spirit of cooperation displayed, all of which tended to reduce somewhat the amount of claims. However, we filed during the past season claims amounting to $71,419.87, of which $33,602.37 has been collected to date. Within the past year we have collected on outstanding claims a total of 4,369, amounting to $93,790.63.

ADVERTISING DEPARTMENT

Exchange advertising is handled by a committe of the board of directors, and has been placed through the Thomas Advertising Service, Jacksonville, Florida, from the beginning.

An outstanding feature of the Florida Citrus Exchange advertising during the year was the degree to which it was made to cover a larger portion of the country than in any preceding season.

SEALD SWEET advertising appeared in more than 25,000,000 copies of nine of the leading home magazines of the United States—full color pages were printed in over 6,000,000 copies.

Eight of the leading and most influential medical and nursing journals of the country carried SEALD SWEET messages in about half a million copies. In addition, there was the usual space in fruit trade and other similar periodicals.

Between November and March, SEALD SWEET advertisements were inserted in approximately 60,000,000 copies of 245 leading daily newspapers. Dealers advertising, paid for by themselves, supplemented the Exchange campaign in a considerable number of these papers.

Requests for our recipe booklet, due principally to the magazine and newspaper advertising, have been much more numerous than previously. The second edition of 100,000 copies was exhausted before the end of the season, though the first edition of the same size lasted nearly two years.

COOPERATION SECURED

The cooperation of most of the factors concerned in making our advertising a success was notably good this year.

To a greater extent than ever before, our sales representatives in the North took advantage of the campaign in an efficient and persistent way.

Fruit dealers, too, gave increased assistance, including, as noted above, the purchase of a large volume of space at their own expense.

Some magazines sent out handsome circular matter to the fruit trade, calling attention to Exchange copy.

Other periodicals had salaried representatives call on leading wholesale houses and explain the extent of our campaign, urging them to handle SEALD SWEET.

Perhaps of even greater value, however, has been the cooperation given by the magazines through their editorial departments, in featuring special articles devoted to citrus fruit, especially grapefruit.

The service rendered by newspapers in helping to make our work effective was on a broader scale than in preceding years. The publishers of the dailies in scores of cities assisted in getting both wholesale and retail dealers to place SEALD SWEET grapefruit and oranges for sale.

Many of the papers went further, inducing the dealers to pay for advertising of SEALD SWEET fruits themselves, in addition to our own. Almost without exception, the newspapers on our list gave material aid in the distribution of our store cards and other illustrated advertising matter. It is safe to say that if the work done for us by the service departments of the newspapers had been paid for, it would have cost the Exchange anywhere from $50,000 to $100,000.

The fact that it has been possible, in the past year or two, to extend the scope of our magazine and trade journal advertising, unquestionably has made it easier for the sales department to open new markets for SEALD SWEET grapefruit and oranges, and to increase the number of dealers pushing our brands of fruit. Practically the entire trade desires newspaper advertising in addition, however, and the difficulties of holding our territory have been multiplied wherever the Exchange was unable to advertise in the newspaper.

It is significant that this year we failed to make sales of SEALD SWEET fruit in less than a half dozen cities and towns in which newspaper advertising was carried during the 1921-22 season. A much larger number of the places which bought in carload lots last year, but in which we were able to do no advertising either last season or this, failed to buy our fruit during the current year. In making plans for next season's advertising, it is our intention to provide campaigns in the largest possible number of the new markets opened this season.

Results Obtained

In response to questions asked them about the close of the shipping season, 61 out of the 72 Exchange district and division manager workers stated that the advertising had helped them to sell more SEALD SWEET grapefruit and oranges than they could have sold without it.

At the same time, 1,256 out of 1,463 representative fruit dealers, wholesale and retail, in 121 cities, testified that they found Exchange advertising helpful in increasing demand for SEALD SWEET fruits. Twelve hundred and thirty-two of these dealers said they could have done an even larger business in SEALD SWEET brands if there had been more advertising.

The great expansion in consumer demand of Florida grapefruit, undoubtedly due to Exchange advertising, is worthy of special stress. When the first aggressive campaign for grapefruit was undertaken by us, our advertising agency made a survey, on which was based the estimate that not more than 5 per cent of the people of the United States ate grapefruit in any quantities. A similar survey, made during the present year, indicates it is reasonably safe to conclude that now 15 per cent of our population eat grapefruit at more or less regular intervals.

In an article published by the *American Magazine* for September, 1922, descriptive of the experiences of a woman who runs restaurants in Chicago feeding 4,000,000 people a year, the author quotes this lady, Miss Mary L. Dutton, as saying, "The year round, grapefruit is the leader among uncooked fruits. We serve it three times a day."

Several national advertisers of food products, including *H. O.* the breakfast food, have featured grapefruit in the recipes they publish in magazines. There have been constantly increasing calls for our literature from domestic science schools and similar institutions, and our workers in this field have been warmly welcomed.

Coloring Fruit

There has been a difference of opinion among growers regarding the advisability of the practice of coloring fruit. Most growers have misunderstood this operation and feel that the SEALD SWEET trademark is liable to suffer by this practice. With this in mind your manager wishes to make this explanation.

Some two years ago we were informed by the United States Government investigators that they had learned that the gas exhaust from a gasoline engine would color fruit. That is, this gas tended to remove green color from the rind and the yellow color was left. When immature fruit was subjected to this process, a very pale lemon color was the result; but on fruit testing above the government requirements for immaturity, a color was obtained like that on the tree.

Government inspectors had not progressed far enough, in their opinion, to warrant them in advising as to installation of coloring plants in various packing houses, but during the latter part of last season we were able to secure a man who had much experience in coloring fruit in California. Although his system was slightly different from that employed by the government inspectors, his credentials were such that we employed him, and last year shipped a considerable amount of fruit, principally Valencias, that were artificially colored.

At the same time the government had a man in the state, working along the same lines as we were working, and by the end of the season we felt that we were ready to go ahead and install rooms in various packing houses. This coloring is progressing satisfactorily, and we are firmly of the opinion that it will prove profitable to the growers. As the system is perfected from time to time, it will no doubt come into general use throughout the State.

Coloring fruit will operate, we believe, to minimize the shipment of immature fruit, because it will be but a short time until the trade will accept only full-colored fruit from all Florida shippers. Although the government is keenly interested in developing this process, it will not permit shipments that have been artificially colored, unless the fruit passes the legal test for maturity, commonly known as 7 to 1 for grapefruit, and 8 to 1 for oranges. The bureau of chemistry has sufficient power, under federal laws, to confiscate any shipment if the fruit is artificially colored and immature.

This will probably operate to make the shipment of green fruit, uncolored, unprofitable, and make an illegal shipment subject to seizure anywhere in the United States.

Canned Grapefruit

Canning of grapefruit is still in the experimental stage. However, we have to report some progress. After this office had suggested a plan whereby a purely cooperative association might be organized for the canning of grapefruit, six associations in Polk County became keenly interested, forming a cooperative canning association. Under the proposed operation of this association, the grower will receive all the money that his fruit brings in the market, less the cost of canning. As the association will operate at cost under exactly the same plans as an association packing house, the grower is assured of no intervening profit between himself and the wholesaler. If this venture is successful, the grower will be in exactly the same position as he is today with his shipments of fruit through the packing house.

Sales will be made through the sales department of the Exchange, and the association will operate under a season pool, beginning about December 1. They have arranged to purchase adequate machinery, and propose to put out a high grade product under a brand that will be controlled by the Florida Citrus Exchange, in exactly the same way as SEALD SWEET. Reports will be made during the season regarding progress.

PRECOOLING

Mr. J. W. Andrews, a former government employee, has compiled all records, and has been instrumental in working out all problems in this work. He has held this position for two years. Credit must also be given to the splendid cooperation of the association managers and their assistants. Also, we wish to acknowledge the enthusiastic assistance, and expert advice, of Paul Mandeville, of the Davenport Sales Company, and Mr. George Braungart, of the Southern Construction Company. The untiring efforts of these gentlemen, in working out the problems of our first coolers, brought about the speedy results in efficiency obtained.

Precooling has received considerable attention during the past year. Three new plants have been erected, making a total of seven plants at Exchange houses. Refrigeration research work has been continued and this work, with comparative data, is proving valuable to member associations.

One extremely important advancement in precooler design and operation was accomplished in the 15-car vegetable cooler erected at Sanford, by the Sanford Farmers Exchange. After careful preliminary tests it became apparent that much time could be saved in cooling such vegetables as celery, lettuce and peppers, if ice-water were used as the cooling medium, instead of cold air methods.

It was found that the ice-water system also produced marked improvement in the quality, as the vegetables were not wilted during the cooling process. Instead, they become so charged with the cold water that they remained fresh and crisp. These investigations led the Sanford growers to erect a plant in which their produce could be washed, packed, cooled, loaded into cars, and cars supplied with ice. The plant, costing approximately $110,000, was erected in 73 working days, and with materials bought almost entirely through the Exchange Supply Company.

A general idea of advancement along cooling lines developed at this plant, can be judged by the fact that a carload of celery can be cooled and loaded in one hour and twenty minutes, whereas the usual time for the cooling alone, under the air circulation system, has usually been from 12 to 20 hours, with a much more costly plant and with two unnecessary handlings of the product. The Sanford plant operated successfully during the season, and the prices obtained on competitive markets of the North for the washed and precooled stock, when reduced to an f. o. b. basis, ranged from 10 per cent to 25 per cent higher than for field packed and standard refrigerated stock from the same locality.

Exchange associations at Kissimmee and Mount Dora are completing new precoolers which will be available for the coming season. In both cases radical changes have been made as compared with plants formerly constructed, and decided economies in construction and operation are indicated. The plants have been designed for rapid, even cooling and, for the reason that no direct comparisons can be made, it would be futile to express here the ultimate advancement hoped for. However, prominent refrigeration engineers, whose criticisms have been solicited, express the opinion that the Exchange is contributing appreciably to the art of precooling and that the new plants will be second to none.

Investigations along precooling lines have been diligently followed for the past two seasons. The first season's work demanded technical tests, and eventually radical changes in houses already in operation, to better adapt the plants to our needs. This work has been continued, but in a lesser degree during the past season. While it is not claimed that perfection has been reached in design and operation, it is evident that precooling is well enough understood, and its basic principles are well enough defined, to warrant investment by such associations as desire to share in its benefits. This view of the subject has occasioned the gathering of statistical data on shipments from all Exchange houses, so that we may be better able to render service to such associations as are contemplating precooling.

Membership Drive

Our membership campaign culminated in an intensive drive during the week of September 11 to 16.

We realize that this year our association and sub-exchange meetings were a little late, as many growers had left the state for vacations and others who only maintained a winter residence here had gone North. Nevertheless, meetings were held in each sub-exchange, at which were present all association directors and association managers, together with the directors of the sub-exchanges. After this series of meetings each sub-exchange arranged for a meeting at practically every association in its territory.

The response of the growers was beyond our expectations, and the realization of the duties of directors and members, when properly explained, created an enthusiasm that was most gratifying.

This organization is built on membership, and so this was termed a membership campaign. We gained many new members and a very considerable volume of fruit. One sub-exchange shows an increase of nearly 50 per cent, not counting fruit that had been purchased in their territory by the Standard Growers Exchange. The gain in membership, and fruit, varied considerably in different sub-exchanges, but there was no difference in the enthusiasm created.

This is probably the first of many similar campaigns, and we feel assured that we have accomplished the first step, which is to bring our present membership into a more solid unit. They now realize that during the past few years they have been accepting many statements as facts that were not facts but that originated from the ambitions of competitors to discredit our organization and its accomplishments.

If this campaign accomplishes nothing more than to have the membership realize that its sole source of reliable information is from the organization itself, we should be satisfied; and we are satisfied, because members now understand that there are no operations about which they cannot get full information, and that it is their duty to keep in close contact with the associations and the sub-exchanges; and further, that it is their duty to spread reliable information regarding the operations of the Florida Citrus Exchange.

CHAPTER 9

1922-1923

THE BOARD of directors for the 1922-23 season was seated at the board meeting of June 8, 1922, and included L. M. Hammel of DeSoto County Citrus Sub-Exchange, D. C. Gillett of Hillsborough County Citrus Sub-Exchange, Henry C. Merrill of Indian River Citrus Sub-Exchange, J. W. Ponder of Manatee County Citrus Sub-Exchange, P. C. Peters of Orange County Citrus Sub-Exchange, R. J. Kepler, Jr., of Volusia County Citrus Sub-Exchange, John A. Snively of Polk County Citrus Sub-Exchange, F. C. W. Kramer, Jr., of Highland Citrus Sub-Exchange, Joy Heck of Dade County Citrus Sub-Exchange, and J. C. Stubbs of Lee County Citrus Sub-Exchange.

It is to be noted with interest that Dr. J. H. Ross, who had until this time served as president of the Exchange and as director representing Polk County Citrus Sub-Exchange for twelve years, was not seated as a member of the board of directors at the June 8, 1922, meeting, being replaced as the Polk representative by John A. Snively. This meeting did, however, see the board approve an amendment of the charter of the Florida Citrus Exchange, which spelled out certain changes in the qualifications of the officers of the cooperative. With regard to the president, the amendment specified that the president need not be a member of the board, provided he qualified as a member of any association affiliated with the Exchange. In the event a board member representing a sub-exchange was elected to the presidency, the amendment required that the sub-exchange from which the new president was elected immediately name a new representative to the board. It seems to have been the intention of the board to place its president above the somewhat embarrassing predicament of exercising judgment for the good of the entire organization while still faced with the responsibility of doing his utmost for the good of his own sub-exchange.

At any rate, Dr. J. H. Ross was re-elected as president of the Florida Citrus Exchange on June 8, 1922, and continued to exercise great influence on the policy of the organization. D. C. Gillett was named first vice-president, and P. C. Peters was named second vice-president. C. E. Stewart was continued as secretary, Business Manager W. T. Covode was continued as cashier, and William Hunter was continued as attorney. Two associate directors were named at this meeting. They were S. C. Warner of East Palatka and E. B. Casler of Pinellas County. In addition to the two associate directors, V. B. Newton was named as special director by virtue of his association as treasurer of the recently merged Standard Growers Exchange, which was now integrated with the Florida Citrus Exchange.

The minutes of the July 7 meeting would seem to indicate that the board's advertising committee and its advertising agency entered upon their first major disagreement in the expenditure of funds. The Thomas Advertising Service, which had handled the Exchange account from the very first, made a strong and compelling argument against the use of streetcar advertisements. Their lengthy presentation outlined agency feeling that money should be used, instead, in additional painted signs for New York, Philadelphia, Chicago, and Boston. To quote directly from the minutes: "The subject under consideration was discussed at length by members of the committee, and on motion duly made and seconded, the committee voted to recommend to the board that the sum of $15,000 be authorized for streetcar advertising."

But the committee's recommendation was not accepted by the board, which apparently could not muster agreement either for or against the streetcar advertisement proposal. As so often happened, the board finally tabled the proposal. Some indication of the deep feeling concerning the streetcar ad proposal is evident in the fact that the matter was tabled at the July 7 meeting, tabled at the September meeting, and tabled at the October meeting, with agreement that it would not again be brought before the board during the current year (1922).

On October 5, 1922, the board took note of the death of M. E. Gillett, first general manager of the Exchange, and one of the original factors in the organization of the cooperative. In a resolution expressing sympathy to Mrs. Gillett, and to their son, D. C. Gillett, at this time a member of the board of directors, the Florida Citrus Exchange paid tribute to the great cooperative work accomplished by the elder Gillett during his lifetime.

Also at the October 5, 1922, meeting, the board finalized arrangements with a group of Polk County associations for the purpose of canning grapefruit. The product would be marketed by the sales department of the Florida Citrus Exchange under the brand-name SEALD HEART, the brand name being made available to any Exchange affiliate which would conform to quality standards established by the organization.

Under an arrangement with the Winter Haven Fruit Products Association, the Exchange agreed to rent that association a packinghouse at Eagle Lake, which formerly had belonged to Standard Growers Exchange, to be used as a factory for canning grapefruit. Although it is possible that earlier canning ventures had been entered into on a much smaller scale, this transaction on October 5, 1922, was the first canning operation undertaken by the Exchange of sufficient importance to have been recorded in the minutes of the cooperative.

Along with the decision to enter the processing field, the Exchange felt it imperative to expand the activities of its chemist, S. S. Walker, to cover further experiments in citrus canning as well as to conduct quality tests to assure high market acceptability of all canned products under the SEALD HEART label. But, with all the apparent progress of the large cooperative, the minutes of this period indicate that the bulky channels of communications between packers and the central office were under constant criticism from the associations.

Typical of this difficulty is an item in the minutes of the December 21, 1922, board meeting:

> The President brought to the attention of the Board that very heavy shipments of citrus fruit had gone forward in the last two weeks, notwithstanding the fact that the sales manager had repeatedly requested the sub-exchange managers to slow up shipments and upon motion of J. E. Klock, seconded by C. J. Stubbs, and duly carried, it was moved that the president and business manager be

authorized to investigate these shipments and ascertain who had disregarded the advice of the sales department.

Apparently in conflict with the nature of this item is another item from the same meeting:

At the last meeting of the board the business manager and attorney were instructed to prepare a bylaw to go in the Specimen By-Laws for Associations. The business manager submitted the following:

Each member of the association shall have the right to control the selection of markets, and the time of shipment of his fruit.

On motion of Mr. Snively, seconded by Mr. Stubbs and unanimously carried, it was voted to adopt the above paragraph for our Specimen By-Laws for Associations.

Another disturbing factor affecting not only the Exchange but the entire industry at this time was the introduction of the blackfly into Florida. K. E. Bragdon of the horticultural department of the Exchange Supply Company appeared before the board and spoke at length on the danger of the introduction of the blackfly into Florida through the port at Key West. Bragdon's address moved the board to notify the State Plant Board that the Exchange was in "hearty sympathy" with a move to have a fumigation plant erected at Key West by the Federal Horticultural Board, and that the organization would if necessary contribute financial support to hasten the program.

At the beginning of the new year, 1923, the Exchange took up a problem which had been arising with alarming frequency for many months. This was the problem of fruit which did not meet specifications of top quality, but which nonetheless remained to be marketed by the organization.

On March 28, 1923, the advertising committee recommended to the board that the SEALD SWEET trade-mark be retained for the major portion of the fruit marketed through the Exchange, and that a new trade-mark be designed and adopted for that portion of the output of its members that was of superior quality, fruit marketed under each trade-mark to be advertised for exactly what it was.

The board was disposed, however, to postpone action on this matter, turning instead to a complicated plan for the erection and equipage of canning factories under the Exchange system. While no record of progress in this field is filed in the regular minutes of the Exchange, there is substantial evidence that special but technically unofficial meetings were occurring frequently as the future importance of citrus processing became more and more apparent.

One problem that had plagued the Exchange since its inception stemmed from those years in which various grower-members of Exchange associations felt that they were incurring financial losses by not being permitted to sell their own fruit to speculators. A committee report submitted and approved by the board on May 23, 1923, made the following provisions:

> That no sale be consummated without consultation with the association, the sub-exchange, and the Florida Citrus Exchange and have their approval.
>
> After the sale has been approved by the sales manager, the sub-exchange and the association, the grower agrees to pay 25 per cent of the combined selling contingency and advertising charges of the Florida Citrus Exchange, 50 per cent of the sub-exchange charge, and 50 per cent of the estimated over-head charge and all of the per-box building and equipment retain of the association.

The committee, composed of F. C. W. Kramer, Jr., John A. Snively, P. C. Peters, George A. Scott, and C. E. Stewart, Jr., preceded these recommendations with the following preamble:

> Realizing that the growers of our organization are being constantly approached by the speculative buyers regarding the purchase of their crops, and realizing that sometimes these cash offers are at first very attractive to the grower because of the cash in hand available immediately, and second, because at times these offers appear to be higher than market warrants or would apparently justify in the immediate future, your committee discussed for some time the possibility of suggesting some plan to you that would permit the grower to take advantage of these offers, but at the same time support his association, sub-exchange, and the Florida Citrus Exchange, and thus still remain a member in good standing, feeling that at times growers felt compelled to accept these offers and under present conditions were placed in a position of breaking their contract or sustaining what appeared to them as a possible financial loss.

The report of the committee was accepted by the board, and the minutes bear a pencilled notation that the report was adopted. At any rate, it was ordered that copies of the report be sent to all sub-exchanges.

The 1922-23 season concluded with the annual meeting which was held on June 5, 1923. The first action of the board was to approve a record expenditure for advertising for the coming season in the amount of $240,000 for fresh-fruit advertising and an additional $20,000 for advertising the relatively new canned grapefruit SEALD HEART.

Credentials of the new board were presented and members were seated for the 1923-24 season. They were John A. Snively of Polk County, E. P. Livermore of Dade County, L. W. Tilden of Orange County, C. J. Stubbs of Lee County, A. F. Wyman of Manatee County, J. W. Perkins of Volusia County, D. C. Gillett of Hillsborough County, W. R. Lee of Marion County, H. G. Putnam of Indian River County, F. C. W. Kramer, Jr., of Highland County, and L. M. Hammel of DeSoto County. V. B. Newton was subsequently seated on the board representing Standard Growers Exchange.

Dr. J. H. Ross was re-elected to the presidency. D. C. Gillett was selected as first vice-president and F. C. W. Kramer, Jr., was chosen as second vice-president. C. E. Stewart, Jr., was designated secretary and business manager, W. T. Covode was named cashier, and George A. Scott was retained in the position of general sales manager.

CHAPTER 10

1923-1924

ONE OF the first matters of business conducted by the 1923-24 board was that of appropriating money for increased advertising of SEALD HEART canned grapefruit. A budget item of $20,000 had been originally scheduled for this purpose, but action of the new board on July 26 increased this amount to $25,000. Also, in regard to advertising, the board agreed on September 19, 1923, to purchase a truck painted with the advertisements of the Florida Citrus Exchange and manned with Exchange salesmen for the purpose of covering the State of Florida to demonstrate SEALD HEART canned grapefruit.

It was during this era that the Florida Citrus Exchange felt compelled to establish its own inspection department on a much larger scale than had heretofore been employed. The decision included dividing the citrus growing area of the state into five districts in such a manner as to make for efficiency in handling and economy in traveling. The plan would include appointment of a chief inspector along with several assistants.

Official action in the employment of the chief inspector finally took place on December 19, 1923, after several weeks of controversy within the organization regarding the man to be employed for the position. Eventually hired for the position was Harold Crews who had for some years been employed as manager of the DeSoto County Sub-Exchange.

Whether the detailed inspection program as envisioned by the board at that time actually fulfilled the expectations of the board is not apparent in records covering this period; however, it must be noted that the establishment of this facility was made in an honest effort to improve both the maturity and quality of fresh citrus being shipped to market under the Exchange brand.

The most far-reaching and informal paper available to the writer from the 1923-24 season is the verbatim report of an address delivered to the board by Business Manager C. E. Stewart, Jr., on January 24, 1924. Because Stewart's address is indicative of the times and problems of this period we quote the message in its entirety:

Gentlemen of the Board, if I may be privileged to say before I read this, that to arrive at a conclusion, suggesting a plan that might put the Exchange in the right light throughout the State of Florida, those premises that were laid down were not chosen for the purpose of bringing a conclusion that was desired by me. When I started to analyze the situation I did not know where I was going to land. I am not trying to sell any pet theories; I do not mind if the Board shoots this plan to pieces. This is the way I see it after trying to take into consideration every factor.

Keen interest is being taken by growers and apparently by some marketing organizations throughout the state to establish some satisfactory method whereby the greater majority of the crop could be marketed in a cooperative way.

The question uppermost in the minds, thinking about the proposition is . . . "Why isn't the Exchange given better support?"

Hundreds of ideas to improve conditions have been put forth, but most of these have bordered upon what might be termed "patented panaceas," and the purpose of this paper is to bring to your attention that perhaps we are in the exact position of the old lady who searched long and hard for her spectacles to finally find them on her own nose. That there is something the matter we must all agree and your Manager believes that the time has come to discuss the situation frankly and bring out, if possible, the causes of unrest, dissatisfaction and criticism.

No organization has been criticized as severely and as constantly as the Florida Citrus Exchange. Practically all of the criticism of the Exchange has been of the Florida Citrus Exchange only as regards the Tampa organization. These criticisms are caused by misinformation and rumor about things that have been done by the directors of the Florida Citrus Exchange. Plans and decisions are made for the good of the business as the directors see it. But the directors are unable to see a sufficient number of their county people to explain purposes, whereas the opposition or some disgruntled member of this organization makes it his business to spread the news as he sees it and no one is there to defend this office or this board of directors until after tremendous sentiment has been built up and any explanations then are only received as excuses.

I submit that the trouble has been, and is, that there is a constant attempt to harmonize twelve different localities of this state on what is other than the sole purpose for which this institution was organized.

The Florida Citrus Exchange was intended originally to, and should, have a single function. This function is divided into three main activities of which there are five necessary off-shoots or contingent activities.

The function of the Florida Citrus Exchange is to sell fruit and to sell fruit intelligently; there should be three divisions of the operation: first, to obtain the very best information possible as to market conditions, showing what to ship and when to ship it; second, to effect a sale when such fruit is shipped; third, to collect the money and remit it to the shipper. Those are the three main and fundamental functions of this organization. There are five other necessary activities; first, judicious advertising for the promotion of sales; second, a satisfactory traffic department, (a) to assist in the handling and diversion of cars, (b) for the collection of claims; third, an inspection department to inspect, (a) the grade and pack of fruit that is put under advertised brands, (b) to assist packing houses in developing proper methods; fourth, a single organizer for the purpose of working in territories that are not developed; fifth, an auditor to see that the money sent the sub-exchange is remitted to the associations promptly and correctly.

All other functions can be handled better by and should be handled by the sub-exchange. Any unrest that has developed in the Florida Citrus Exchange has developed because certain plans, policies or rules were suggested or put in force which were acceptable to certain sub-exchanges but not acceptable to other sub-exchanges.

The citrus belt of Florida is spread over a very large area, and conditions and people are to a considerable extent different in each locality. Sub-exchanges should have the right to order this "ruliness," from an organization standpoint, to meet the conditions as they find them in that locality. The representative sent to this office should consult with the general sales manager as to what is most advantageous to ship, what are the best sizes to ship and in what volume. Throughout the shipping season these meetings should occur not less than every two weeks; these representatives should be sent to Tampa at the expense of the sub-exchange with full authority to enter into agreements with representatives from other sections of the state and the necessary authority to control the production of the packing houses in their territory.

In this office there has developed a certain amount of paternalism. We have helped some sub-exchanges, we have helped some associations, we have a considerable amount on certificates of deposit in various banks throughout the state to help finances of local associations and sub-exchanges. This office needs a certain amount of working capital because we cannot collect enough money in the first part of the shipping season to pay our expenses, and we are always in danger of making expenditures and having our income cut off almost entirely in the midst of the season by some disaster. But the amount of money we need is not considerable. A certain proportion of the funds now held by the Florida Citrus Exchange could be given back to the sub-exchanges so that they could handle their own financial arrangements in their sub-exchange. The one hundred and thirty thousand odd dollars that we have on certificates of deposit is of doubtful benefit to this office but is of considerable benefit to the sub-exchanges. I submit it would be of more benefit to the sub-exchanges if the sub-exchange had the handling of these funds. The sub-exchange must function as a business corporation, closely confined, studying the local conditions constantly and meeting them in a business way, the same as any other commercial marketing or packing company meets them—meet them in the way that the particular territory demands without regard as to how they are met in other sections. This responsibility belongs to the associations and sub-exchanges. Show me a man without responsibilities and I will show you a failure. Show me an unsuccessful association and it is one accepting no responsibilities but depending on the Tampa office and not recognizing its own opportunities. The successful associations, the ones that pay regularly the highest dollars to their growers, are the ones taking on responsibilities.

I know nothing of the operations of the Standard Growers Exchange; they handle their own business, they collect their own fruit, they operate their own packing houses, but our contact with them is solely on the basis of handling the fruit that they ship. This is an ideal condition and if all sub-exchanges contact with this office was confined solely to the handling of the fruit that they ship, you would find that the criticism of the Florida Citrus Exchange, as such, would disappear, that it would function then 100 per cent as an efficient marketing combination where the fruit interest of the entire citrus belt could meet the final decisions and agreements entered into, and by having this fruit go through one channel, it could be seen whether the agreements were lived up to.

There has been too much attention given to influencing the grower in some other way than by paying him the high dollar. Growers in sub-exchanges, instead of going to their headquarters, that is their sub-exchanges, deem it a privilege to come to this office feeling that they are getting closer to the organization. A grower should have only one thought in mind in joining an association; what to ship and when to ship it. All other arrangements with his associations are to help make the association function properly. I know that there are certain rules and regulations that have been put in force that sub-exchanges feel hamper them and other sub-exchanges feel are of an advantage, and if two sub-exchanges operate on slightly different plans, locally, you would find that this would not affect the orderly marketing of their fruit or the establishment of certain brands.

It must be recognized that there is always going to be a certain amount of keen competition between sub-exchanges, and in sub-exchanges between associations. It is true that some associations return a higher price than others to their growers for the same quality of fruit. This shows more efficient management in the local association. This should be built up by the sub-exchange rather than hampered by general agreements entered into a half-hearted way by all associations throughout the entire state. There are more ways than one of accomplishing almost any business enterprise, but we do agree that there is only one way to market the entire crop for the State of Florida and that is that each packing organization should know what the other packing organization is doing and work in harmony with it, under agreements of "what and when to ship."

Everyone realizes that we must have proper advertising to promote the sales of fruit of an organization. It is useless to advertise nationally unless it is what is termed "trademark advertising." Each sub-exchange should pay its proportion on a box basis to support such advertising; but if an association does not wish to conform to the standard set for that advertised brand it should have the privilege of shipping under other brands, but it still should pay its share of the national advertising for the general promotion of sales of the entire crop. This is best adjusted on a box basis. It would probably be best to incorporate the advertising in the general budget and make a flat selling charge to cover it all. All shippers must pay the selling charge whether they use the advertised brands or not.

Recently some associates of managers of one sub-exchange made a trip to some Northern markets. They came back with well-defined ideas to as what was best for their association and what their particular customers desired. Surely if they coordinated the movement of their fruit with the general movement of the fruit through the Exchange they should be permitted to so order their business as seems most advantageous to them from investigation that they have made.

The independent marketing organizations are taking some interest in cooperative marketing. What the commercial marketing organization is considering is losing its own identity, that is, John Doe and Company might be willing to cooperate in the marketing of their crops provided they could still remain John Doe and Company. We see no reason why a commercial marketing organization really desiring to cooperate could not join the Florida Citrus Exchange and operate as a sub-exchange member. If Chase and Company were to join the Florida Citrus Exchange why could not Chase and Company operate as a sub-exchange? They have packing houses, they can buy fruit or they can pack fruit from the groves that they own or have leased. This could be presented through the Florida Citrus Exchange just the same as fruit of any sub-exchange. This would bring about some competition in our sub-exchange territories, but I submit that if a commercial marketing organization can give the growers better service, can operate their packing houses more economically, can get the fruit off with great dispatch, when it is profitable to ship it, that such competition would have a most stimulating effect upon the operation of our associations and sub-exchanges. It would bring out most forcibly that the thing that is holding the grower is the high dollar and service.

Associations that have been outstanding successes in the Florida Citrus Exchange could just as well operate alone and independent under the present disorderly movement of the fruit from the state. The only attraction to those associations in affiliating with the Florida Citrus Exchange is to coordinate the movement of all fruit and establish nationally advertised brands. Therefore, they should be permitted to order their own business in cooperation with their sub-exchange, as seems most advantageous, providing that they cooperate 100 per cent in the orderly movement of fruit.

If these marketing organizations wish to hold their own brands, they can do it of course. They could also put on the boxes the trademark of our advertised brand; they would surely have to pay their advertising assessment, and if they want to take advantage of the national advertising it would be up to them to conform to the grading rules for that advertised brand.

Many growers throughout the state feel that the Florida Citrus Exchange is not willing to make it possible for these commercial marketing organizations, especially those that own and control groves, to work with them, and I suggest that the Florida Citrus Exchange immediately advertise in full-page advertisements, one issue, in all daily papers of this state, the fact that a commercial marketing organization can affiliate with the Florida Citrus Exchange and still retain its identity; that it will have representation on a box basis of the Board of Directors of the Florida Citrus Exchange, the Board's function being solely on questions of selling; and the control of their business will be absolutely in their hands; that the control of business in every sub-exchange will be absolutely in its own hands. Operated in this way, we would have a real representative marketing organization. I believe this advertisement could be made clear enough and forceful enough to convince every grower in the state that the next move was up to the commercial marketing organizations; that we had outlined a plan that was reasonable and workable. The Exchange would be relieved of the burden of criticism and we would prove to every grower of the State of Florida that we were ready and willing to re-vamp this organization to meet the present needs of marketing the crop. The only thing would be that fruit would be presented through the Florida Citrus Exchange and agreements would be entered into as to the volume to be shipped and when shipped through a central marketing organization. It is possible that the commercial marketing organizations would not find it is easy to solicit fruit to handle on brokerage, but if that was the point that defeated this combination the growers of the state would know it and would realize that the independent marketing interests were defeating the establishment of a large cooperative selling organization for purely personal reasons.

I submit that we do not need any "patent panaceas," that we have got in our organization the fundamentals, if rigidly adhered to, of establishing the kind of organization this industry needs; that now is the time to clear the decks and confine our activities solely to the purpose for which we were organized, that of selling fruit, and leave all of the policies and plans calculated to create sentiment and influence growers in the various localities through the state to the sub-exchanges.

I sincerely trust that a committee will be appointed to prepare and pass on a copy of an advertisement to suggest to the state that those marketing organizations that wish to affiliate with the Florida Citrus Exchange can affiliate on the same basis as the sub-exchange; that immediate steps be taken to confine the operations of this office solely to the marketing of fruit; and I assure you that if the operations of this office are gone over carefully with your manager we can reduce expenses, overheads, get rid of unnecessary work and help establish sub-exchanges that will function as real business institutions, operating in localities along the lines that those localities demand. These suggestions are not based on theory.

The foregoing address was accepted by the board, and work was commenced immediately to make the necessary changes and provisions of charter and bylaws to accomplish (a) a change in representation of sub-exchanges to put the same upon a box basis, (b) a change in the operations of the Florida Citrus Exchange which would confine it exclusively to the marketing of the products of its members, and (c) that a commercial marketing organization could affiliate with the Florida Citrus Exchange and operate as a sub-exchange at large.

To the contemporary citrus man, it will become at once apparent that most of the recommendations of Stewart have come to pass and that most of his philosophy concerning the Exchange remains today as a part of the organization's policy with regard to its members.

It should also be noted in this chapter that three new employees of the Florida Citrus Exchange during the 1923-24 season were Charles Felix, Frederick Swain Johnston, and E. F. Gudgen, all of whom were still employed by the firm at the close of 1959. Johnston, who retired on January 6, 1960, completed his long career as general sales manager. Felix is an assistant sales manager, and Gudgen holds the position of auditor for the Exchange.

CHAPTER 11

1924-1925

WHILE the preceeding chapter will suffice to tell much about the 1923-24 season, this history must record that during the latter months of that season, Dr. J. H. Ross, who had served as president of the Florida Citrus Exchange during the turbulent years since the 1914-15 season, decided to resign from active participation in the affairs of the Exchange.

The resignation of Dr. Ross was deplored by most members of the board, who felt that the leadership of the aging Ross was indispensable to the Exchange during these times. That he was a leader there can be no doubt. He was, on the basis of records still on file at the Exchange, a prolific writer and an effective speaker. A firm believer in the cooperative movement, Ross carried the Exchange banner high. He was unyielding in his search for methods to improve the quality of Florida citrus, and time and time again he badgered, berated, and pleaded with the Florida citrus industry to change its ways—to improve quality, and to cooperate to a fuller extent in the sale of Florida citrus. There can be little doubt that the policies of the Exchange reflected Dr. Ross's dynamic personality. His ideas, if considered in the light of what was to be the future of the Florida citrus industry, were well ahead of the times. His white hair and flowing white mustache, a familiar sight at almost any citrus meeting, framed an intelligent countenance known and respected throughout the industry.

Thus at its first meeting, the board just elected for the 1924-25 season received and accepted with regret the resignation of Dr. Ross on June 5, 1924. The board consisted of E. P. Livermore of Dade County Citrus Sub-Exchange, L. M. Hammell of DeSoto County Citrus Sub-Exchange, John A. Snively of Florence Villa Citrus Sub-Exchange, F. C. W. Kramer of Highland Citrus Sub-Exchange, D. C. Gillett of Hillsborough Citrus Sub-Exchange, Homer Needles of Indian River Citrus Sub-Exchange, R. O. Philpot of Lake Region Citrus Sub-Exchange, Edward Parkinson of Lee County Citrus Sub-Exchange, Josiah Varn of Manatee County Citrus Sub-Exchange, Walter R. Lee of Marion County Citrus Sub-Exchange, J. S. Cadel of Orange County Citrus Sub-Exchange, John S. Taylor of Pinellas Citrus Sub-Exchange, Vet L. Brown of Polk County Citrus Sub-Exchange, E. L. Wirt of Ridge Citrus Sub-Exchange, A. V. Anderson of Scenic Citrus Sub-Exchange, R. J. Kepler, Jr., of Volusia County Citrus Sub-Exchange, H. E. Cornell of Winter Haven Citrus Sub-Exchange, W. W. Yothers of Lake Apopka Citrus Sub-Exchange, L. W. Tilden of Seminole-Orange Citrus Sub-Exchange, and W. E. Lee and V. B. Newton as special directors.

With regard to the resignation of Dr. Ross, a motion was placed before the board that Dr. Ross be appointed president-emeritus of the Florida Citrus Exchange and that his salary be continued for the balance of his lifetime. To this motion, Dr. Ross made the following reply:

I would like to say one word if the board will indulge me. I find myself now in a very embarrassing position. I am telling you frankly that I would not accept this honor if you voted it with the understanding that my salary continue. I am not a rich man but fortunately I am not a pauper. I do not need the salary. I have always considered it as purely nominal. Now that more or less active service will be discontinued I certainly could not, and I think it would be very unwise, for this board to offer me a salary and I could not possibly accept it. The other side of the motion, a man would be less than human if he did not appreciate the expression of your confidence by electing him emeritus president of the organization. If that came to me, and I thought it was unanimous, a real expression of your feelings, I should place it among the great many hundreds of things that have been said to me that would be a comfort to me during the balance of my life. It would be purely an expression of the confidence and esteem you had for me.

With this brief address, Dr. Ross retired from the presidency of the Florida Citrus Exchange. He was unanimously elected to the position of president-emeritus and retired in high respect by the great majority of all those who knew him. Subsequent action of the board also bestowed upon Dr. Ross the degree of Doctor of Cooperative Service.

The board then undertook the task of electing a new president. In short order, the names of Dr. W. A. Mackenzie of Leesburg and L. C. Edwards of Thonatasassa were placed in nomination. On the first count from a total of 19 votes, Edwards received 12 votes and Mackenzie received 7. A second roll call vote saw the unanimous election of Mr. Edwards to the presidency.

Other officers elected at this time were F. C. W. Kramer, Jr., as first vice-president, John A. Snively as second vice-president, and E. L. Wirt as chairman of the board. Elected as secretary was C. E. Stewart, Jr.; and W. T. Covode was elected to continue as the cashier of the organization. Judge William Hunter was continued as the legal counsel for the Exchange, the general sales manager was George A. Scott, and E. D. Dow was continued as traffic manager.

Two additional vice-presidents were added to the officers of the board on June 18. They were L. W. Tilden as third vice-president and V. B. Newton as fourth vice-president.

Several matters indicative of the times and of major importance occurred during the 1924-25 season. Among these was a decision by the board to seek for and employ a full-time advertising manager capable of administering the Exchange's advertising program, which was expected to reach above $400,000 for the 1924-25 season. While no immediate action was taken on this matter, the minutes of the August 20 meeting indicate the employment of John Moscrip as a full-time advertising director. The records of the July 16, 1924, meeting also indicate the employment of W. E. Lee as chief organizer for the organization.

A resolution emerging from this meeting is particularly interesting because it foreshadowed the course of things to come.

WHEREAS, in 1909 the North DeSoto Citrus Sub-Exchange, also the Arcadia Citrus Sub-Exchange, were lawfully incorporated under the laws of Florida and both were duly recognized and admitted as Sub-Exchanges in the Florida Citrus Exchange, and,

WHEREAS, afterwards, in 1910, the said two sub-exchanges attempted by resolution passed in joint meeting of their boards of directors to consolidate and operate under the name of DeSoto Citrus Sub-Exchange but were never chartered under the laws of Florida, and,

WHEREAS, the Board of Directors of said DeSoto Citrus Sub-Exchange has decided to dissolve said consolidation, and,

WHEREAS, the North DeSoto Citrus Sub-Exchange has amended its charter so as to change the name to the Hardee Citrus Sub-Exchange,

THEREFORE BE IT RESOLVED, that this board of directors recognize the said Hardee Citrus Sub-Exchange, covering all of Hardee County and comprised at present of

Limestone Citrus Growers Association
Ona Citrus Growers Association
Wauchula Citrus Growers Association
Zolfo Citrus Growers Association

and the Arcadia Citrus Sub-Exchange, embracing the territory in DeSoto and Charlotte counties and at present comprised of

Arcadia Citrus Growers Association
Fort Ogden Citrus Growers Association
Nocatee Citrus Growers Association
Punta Gorda Citrus Growers Association
Russ and Hollinsworth of Brownville
B. F. Welles of Fort Ogden

and that the directors elected by each of said citrus sub-exchanges be seated as members of the Board of Directors of the Florida Citrus Exchange.

The resolution was accepted by the board and subsequent action saw L. M. Hammel of the defunct DeSoto County Citrus Sub-Exchange reseated as a director representing the Hardee Citrus Sub-Exchange. The director representing the Arcadia Citrus Sub-Exchange was B. F. Welles of Arcadia. The name of the Arcadia Sub-Exchange was changed later in the year to the Charlotte Citrus Sub-Exchange.

It is interesting to note that the advertising budget for this season was based on an expected 10,000,000 boxes of citrus to be handled by the Exchange at an advertising assessment of four cents per box.

The expenditure of these funds was broken down to include salaries at $25,000, art work and engravings at $20,000, literature (display racks, store cards, etc.) at $10,000, medical and nursing journals at $4,000, magazines at $100,000, newspapers at $133,000, billboards, streetcars, exhibits at $25,000, and sales service at $25,-000. A miscellaneous category, along with a special bulk-fruit advertising program would take up the remainder of the $400,000 budget.

The matter of billboard and streetcar advertising which had traditionally sparked controversy among members of the board again brought forth considerable argument. To quote Josiah Varn:

I have been much interested in the outcome of the billboard advertising into which the committee was requested to look. The whole country is now fighting the billboard advertising to the extent of boycotting people that are advertising on billboards.

C. E. Stewart, business manager, asserted:

I will say that the advertising committee has taken the matter up with various cities. They have instructed me to make investigation over the country as to the sentiment of billboards. We have taken forty of the largest cities and we have written the women's clubs, Rotary Clubs, Kiwanis Clubs, and boards of trade and asked them for their opinion as to the use of billboards not on the highways but within the city limits, and up to the present time no one has objected to the use of billboards in cities. Everybody is in favor of billboards in cities, they are all against billboards on highways. Before that committee will make recommendations they will have a complete report on the attitude of people all over the country on billboards.

The matter of green and immature fruit, for so many years a problem spot for the Exchange, again plagued the board during the 1924-25 season. In an effort to curb shipment of green fruit from the state as a whole, the Exchange appropriated and utilized a $10,000 special advertising fund directed toward all Florida citrus interests asking special consideration of businessmen and growers in the confining of shipments of fruit from Florida to that fruit which would pass the test required by the federal government.

Another recurring problem, that of insect pests, arose during the 1924-25 season and prompted the following resolution:

WHEREAS, it has come to the attention of the Board of Directors of the Florida Citrus Exchange that some interests are endeavoring to change provisions of Quarantine #56, which prohibits the entry into the United States, in and through Florida, of various fruits and vegetables from other countries, and

Whereas, it is the firm belief of the members of this board, who represent upwards to six thousand growers of the State of Florida, that any changes in the above referred to quarantine law would jeopardize the agricultural and horticultural interests of the State of Florida, through the possible introduction of dangerous plant pests,

Therefore, the Board of Directors of the Florida Citrus Exchange, in session, this 20th Day of August, 1924, resolves to respectfully request the Federal Horticultural Board to make no changes in Quarantine #56 which would tend, in any way, to weaken the protection which is now afforded by this quarantine.

This season, 1924-25, was to see the replacement of many Exchange fruit inspectors by the Federal Inspection Service. A resolution passed by the board on August 20, 1924, required that all houses using the SEALD SWEET label accept federal inspection as one of the prerequisites for using that label.

With the continued expansion of the Florida citrus industry, and what was at that time a new consumer trend toward packages, it became inevitable that the Florida Citrus Exchange become involved in the many ramifications of citrus fruit containers. Thus the board, on August 20, 1924, accepted a proposal extended by the FIBOPAK company of Chicago, which would establish within the Exchange a department known as the Consumer Package Division. The packages to be used by this department would be exclusively designed and produced by the FIBOPAK organization. Under the terms of the agreement, the direction and operation of the newly created department would be under FIBOPAK jurisdiction, for which that firm would conduct promotional campaigns, provide a sufficient number of jobbers in each market, employ special dealer service personnel to work with this department, and conduct a special advertising campaign for Exchange fruit packed in the FIBOPAK containers.

Though the board accepted the terms letter of the FIBOPAK organization, there is no further reference to the proposal in the minutes of board meetings during the balance of the 1924-25 season. It must be assumed, therefore, that events in the immediate future had indefinitely postponed action in this matter. But the arrival of new trends in consumer packaging at this relatively early period in citrus history is sufficiently interesting as to command notice in these writings.

Other significant happenings of the 1924-25 season are recorded here briefly with no attempt to conform to a chronological sequence.

On September 17, 1924, the board agreed to subscribe to an insurance policy extended by the Insurance Company of North America, the Automobile Insurance Company, the Fenix Insurance Company of Hartford, the Providence-Washington Insurance Company, and the Camden Insurance Company for the insurance of the income of the Florida Citrus Exchange against loss on account of freeze. The policy was issued in the amount of $450,000.

Also on September 17, 1924, the Florida Citrus Exchange unanimously agreed to embark on its first premium program in the form of juice extractors which would be distributed on a premium basis to consumers throughout the nation.

The Florida Citrus Exchange, also at this meeting of the board, established various grades of citrus fruit to be sold under the SEALD SWEET label. They were SEALD SWEET fancy, brights, goldens, or bronze.

The September 17 meeting saw the Exchange go on record as favoring shipments of citrus by water and as advising the president to take up with each association and explain in detail the arrangements from packing point to destination, asking their cooperation and support for a definite commitment of fruit.

The organization committee, on October 15, 1924, recommended that the Exchange accept the "Dr. Phillips" organization of Orlando as a sub-exchange, a recommendation that was favorably considered by the board in the light of the more than 500,000 boxes of fresh citrus handled annually by the Phillips organization.

The merger of Standard Growers Exchange into the Florida Citrus Exchange during the season preceding was rescinded by action of the board and the merger was dissolved.

Although it had been previously required that all fruit labeled SEALD SWEET be federally inspected, there were some who strongly objected to this. Controversy within the Exchange over the exclusive use of the Federal Inspection Service by all members of the Exchange, rather than on a voluntary basis substituting the Exchange's inspection service where federal service was not used, was resolved in a resolution which would bar utilization of the SEALD SWEET brand on fruit not undergoing federal inspection.

On January 21, 1925, a matter which had been brewing for many months finally came before the board for action. The board's decision was to become one more milestone in the evolution of the organization. That there had been some dissatisfaction with the Thomas Advertising Service, agency of the Exchange since its first venture into advertising in the 1909-10 season, is apparent from the various actions of the board with regard to the administration of its advertising program. While no specific reference to the reasons for the decline of favor is made in the minutes, there is indication that the considerable sums of money being appropriated in this period for advertising purposes left board members with some doubts as to the capability of the Florida agency. This is apparent in the reasons leading up to the employment of a full-time advertising director and subsequent authority placed upon this director. More and more, the advertising business of the Exchange had become a matter of direct negotiation between the Exchange's advertising director and media in a sort of bypass of the Thomas Advertising Service. Thus, on January 21, 1925, the Exchange allowed presentations by three large agencies, finally employing the services of Erwin, Wasey and Company, of Chicago, effective at the beginning of the 1925-26 season.

Another change in the operational procedure was indicated when H. G. Gumprecht, president of the Sub-Exchange Managers' Association, registered the protest of that group in the manner of handling bulk fruit. As the outcome of this protest, the Exchange abolished the bulk fruit sales department and ruled that bulk fruit sales should be handled in the future by the sales department in the regular way.

On April 10, 1925, the Exchange, led by John A. Snively, voted to purchase four cars of grapefruit for an export experiment under the supervision of Sales Manager George Scott.

At the close of this season, the Exchange was in possession of a voluminous report by its new advertising agency with regard to the general operation of the organization and numerous suggestions as to how the firm might revise operations and policy to effect greater efficiency in sales.

Several changes in the composition of the board during the season had occurred before the close of the 1924-25 season. In the closing months, the board consisted of E. L. Wirt, chairman of the board; L. C. Edwards, president; and F. C. W. Kramer, Jr., John A. Snively, L. W. Tilden, Walter R. Lee, L. M. Hammel, Homer Needles, John S. Cadel, B. F. Welles, R. O. Philpot, A. J. Dozier,

Vet L. Brown, H. E. Cornell, Lorenzo A. Wilson, R. J. Kepler, Jr., C. H. Walker, Edward Parkinson, F. S. Ruth, Josiah Varn, A. C. Castle, D. C. Gillett, W. W. Yothers, John S. Taylor, and W. E. Lee.

CHAPTER 12

1925-1926

THE 1925-26 season was to see many changes take place in both the organizational structure and the policy of the Florida Citrus Exchange. These changes were to be mostly the direct result of the extensive survey of Erwin, Wasey and Company mentioned in the preceding chapter. Probably the most far-reaching change, and certainly the most enduring, was the decision to re-establish the office of general manager of the Florida Citrus Exchange. In other matters, the Exchange's board obviously began at this time to flex its muscles in its association with the Florida citrus industry.

Let us first, however, consider the board itself as it was seated at the June 2, 1925, annual meeting. Although the complete board did not form until later in the month, the names of members cited herein represent the board as it eventually took shape later on in the year:

B. F. Stewart of Charlotte County Sub-Exchange, A. E. Barnes of Dade County Citrus Sub-Exchange, A. J. Dozier of DeSoto Citrus Sub-Exchange, R. O. Philpot of Lake Region Citrus Sub-Exchange, H. E. Cornell of Winter Haven Citrus Sub-Exchange, John A. Snively of Florence Villa Sub-Exchange, A. V. Anderson of Scenic Citrus Sub-Exchange, Vet L. Brown of Polk County Citrus Sub-Exchange, L. M. Hammel of Hardee County Citrus Sub-Exchange, F. C. W. Kramer, Jr., of Highland Citrus Sub-Exchange, W. J. Ellsworth of Hillsborough Citrus Sub-Exchange, W. W. Yothers of Lake Apopka

Citrus Sub-Exchange, W. W. Raymond of Lee County Citrus Sub-Exchange, Josiah Varn of Manatee County Citrus Sub-Exchange, R. Lee of Marion County Citrus Sub-Exchange, L. W. Tilden of Seminole-Orange Citrus Sub-Exchange, J. S. Cadel of Orange County Citrus Sub-Exchange, R. J. Kepler of St. Johns River Citrus Sub-Exchange, F. S. Ruth of Mountain Lake Corporation, W. E. Lee, special director; D. C. Gillett, special director, J. S. Taylor of Pinellas Citrus Sub-Exchange, Jay Burns, Jr., of Ridge Citrus Sub-Exchange, H. H. Kellerman of Indian River Citrus Sub-Exchange, and Lorenzo A. Wilson of Hardee Citrus Sub-Exchange.

E. L. Wirt was elected to the presidency and re-elected to the chairmanship of the Florida Citrus Exchange. F. C. W. Kramer, Jr., was re-elected as first vice-president, John A. Snively as second vice-president, L. W. Tilden as third vice-president, and Walter R. Lee as fourth vice-president.

Thus it was that the stage was set on June 2, 1925, for the employment of the Exchange's first general manager in many years. C. E. Stewart, Jr., who had for many years supervised the staff activities of the Exchange as business manager and secretary had tendered his resignation and was no longer in the employ of the organization. After three executive sessions and several caucuses, C. C. Commander, manager of the Polk County Citrus Sub-Exchange, was named to the position of general manager. This action helped shape much of the future of the Exchange. It is apparent from the minutes that Commander, was, from the very first, in complete charge of the administrative activities of the organization. Action taken at the June 2, 1925, and subsequent meetings during this season saw the board placing more and more responsibility in the hands of its new general manager, who, with the support of an executive committee, became virtually unlimited in authority to act on behalf of the board.

O. M. Felix was named as acting secretary of the organization, and subsequent action of the board confirmed his election as secretary. W. T. Covode was again named as cashier, George A. Scott as sales manager, General A. H. Blanding as production manager, E. D. Dow as traffic manager, and John Moscrip as advertising manager. Judge William Hunter was again retained as the Exchange's legal counsel.

One of the first actions of the new board was to reconsider its actions of the previous season with regard to the utilization of federal inspectors. Following a motion by John Snively, on June 17, 1925, that the Exchange use its own inspection service under the supervision of its production department in all packinghouses, the board voted to discontinue federal inspection service for the 1925-26 season. With its executive committee doing spade work on all important matters to come before the board, meetings of the full board during this period were held less frequently and for the most part were confined to either approval or disapproval of recommendations of its executive committee.

In retrospect, it seems almost symbolic that the era of C. C. Commander should have begun only a few days before the citrus industry learned of the death of Dr. J. H. Ross, beloved president-emeritus of the Exchange. To this writer Dr. Ross symbolized the philosophy and hopes and ambitions of the Exchange's first fifteen years of struggle against sometimes hopeless odds, while the new general manager was to become the "Little General" in the Exchange's industrial wars of the new era.

Recorded in the minutes of the March 17, 1926, meeting is the resolution of the Florida Citrus Exchange commemorating the death of Dr. Ross.

Dr. John Harvey Ross

He is gone. All Florida mourns his passing.

And well it may. As friend, counsellor, and builder his place among the foremost leaders of the state cannot soon be filled. But to us, whose privilege it has been to know his companionship, his death brings far greater sorrow. We have felt his magnetic personality, his sincerity of purpose. Business companionship and personal friendship have taught us his sterling character.

Now forever gone from our councils, his absence becomes increasingly significant of the greatness of our loss. Some comfort remains. Though he is no more, the memory of the active man, friend and companion cannot be taken from us. The vital benefits of his work remain manifest. His example is before us, a fitting subject for emulation.

It is far beyond our power, with the feebleness of the spoken word, to express our appreciation of him and his work. Verbal expression is too often trivial, common. It is far more proper that we dedicate ourselves to the perpetuation of the monument that is his achievement.

Let this action, then, be our appreciation, that the ideals toward which he unswervingly strove may be written into an even more complete record of accomplishment.

Thus was commemorated the later life and works of Dr. J. H. Ross, Doctor of Cooperative Services, and president-emeritus of the Florida Citrus Exchange.

On March 17, 1926, the board listened to and endorsed a proposal by C. C. Commander which may well have been the origin of the industry's action in later years to establish the Florida Citrus Commission. While Commander's proposal cannot be closely paralled to the commission's, it incorporates the trend of thought out of which the commission was eventually fashioned. As a representative of the Florida Citrus Exchange and a member of the Fruitmen's Club of Florida, Commander told the Exchange board that he would offer his proposal to the Fruitmen's Club in the form of a motion at its next regularly scheduled meeting.

THE PROPOSAL

1. That an organization be created and incorporated under the laws of Florida for the sole purpose of the financing, formation and direction of publicity and consumer advertising of Florida citrus fruits.

2. All memberships in that organization be equally available to all citrus shippers of the state.

3. Finances for the expressed purposes of the organization be derived by means of a per box retain, the amount of which may be varied from season to season.

4. Voting power in the organization be on a per box basis; that executive power on matters of routine business be vested in a Board of Governors.

5. A subsidiary company be formed for the purpose of developing European markets for the overflow of Florida fruit; that this subsidiary company will include the development and direction of a foreign sales organization to handle fruit furnished by all members on a pro-rata basis.

6. The functioning of this organization will in no way control sales policies of individual shippers or organizations, offer sales resistance, inhibit individual brand advertising or control prices; that its purpose be limited to the direction of publicity.

7. A committee of seven members of the Fruitmen's Club be selected to study this proposal as generalized above and discussed

A group of citrus growers posed for this picture in 1898. They were meeting at the ampa Bay Hotel to discuss the problems of an over-abundant four million box Florida trus crop. It was out of a similar discussion that the Florida Citrus Exchange was formed even years later.

A group of growers, interested in the Exchange, visited the Temple Packing House in 'inter Park early in 1910 to study the most modern facilities available for packing and ipping citrus fruits.

THE "ROARING TWENTIES"

The Exchange promoted its brand name SEALD SWEET at every opportunity, seldom overlooking an occasion on which to combine talent and beauty to boost the sale of Florida citrus.

The Florida Citrus Exchange was an ear subscriber to broadcast advertising, both organizational promotion and in consum product advertising. The photo above shov a typical Exchange broadcast during the 192 21 season.

In 1926, demonstration trucks were covering the United States to promote canned grapefru sections. The truck above was stationed in Louisville.

in detail on succeeding pages, and draft a report to be presented to this body for action at its next regular meeting.

There is indication in the minutes of the April 13, 1926, meeting that the Commander proposal was not accepted in its entirety, but the board did authorize its general manager to accept a substitute proposition issued by the Fruitmen's Club if such a proposition seemed acceptable to him.

The annual meeting of the board for the 1925-26 season occurred on June 1, 1926. Seated as directors for the 1926-27 board were W. O. Talbott of Dade County Citrus Sub-Exchange, A. J. Dozier of DeSoto County Citrus Sub-Exchange, B. F. Stewart of Charlotte Citrus Sub-Exchange, L. M. Hammel of Hardee Citrus Sub-Exchange, F. C. W. Kramer, Jr., of Highland Citrus Sub-Exchange, W. J. Ellsworth of Hillsborough Citrus Sub-Exchange, H. H. Kellerman of Indian River Citrus Sub-Exchange, Josiah Varn of Manatee County Citrus Sub-Exchange, W. R. Lee of Marion County Citrus Sub-Exchange, F. A. Rundle of Orange County Citrus Sub-Exchange, W. W. Yothers of Lake Apopka Citrus Sub-Exchange, H. C. Tilden of Seminole-Orange Citrus Sub-Exchange, John S. Taylor of Pinellas Citrus Sub-Exchange, Vet L. Brown of Polk County Citrus Sub-Exchange, John A. Snively of Florence Villa Citrus Sub-Exchange, R. O. Philpot of Lake Region Citrus Sub-Exchange, Jay Burns, Jr., of Ridge Citrus Sub-Exchange, A. V. Anderson of Scenic Citrus Sub-Exchange, H. E. Cornell of Winter Haven Citrus Sub-Exchange, and R. J. Kepler, Jr., of St. Johns River Citrus Sub-Exchange.

W. E. Lee and F. S. Ruth were subsequently seated as special directors as was D. C. Gillett.

Again elected to the presidency and as chairman of the board was E. L. Wirt. O. M. Felix was re-elected as secretary, F. C. W. Kramer, Jr., as first vice-president, John A. Snively, as second vice-president, H. C. Tilden, as third vice-president, and W. R. Lee, as fourth vice-president. C. C. Commander was named general manager, George A. Scott, sales manager; General A. H. Blanding, production manager; E. D. Dow, traffic manager; John Moscrip, advertising manager; W. T. Covode, cashier; and Judge William Hunter, legal counsel.

CHAPTER 13

1926-1927

THE 1926-27 season was a good one for the Florida Citrus Exchange. By the end of the season it had increased its percentage of control of the total Florida crop by purely voluntary methods as Exchange results in favorable sales continued to mount in spite of serious freezes on January 12 and on January 16, 1927. These freezes, along with generally cold weather and extremely high winds, forced the industry to its knees and more than one citrus forecaster saw almost complete breakdown of the industry.

As has been the case until fairly recently, frozen fruit was shipped to market in large quantities. In the words of C. C. Commander, "Customers who used to swear by Florida citrus are now swearing at us." Nonetheless, by May 10, 1927, the Florida Citrus Exchange had shipped 12,417 cars of citrus to market, a volume that was estimated at slightly over 30 per cent of the total Florida crop. And up until that date, the Exchange had returned to its associations more than $9,000,000 and predicted that this total would be increased by another $1,500,000 before the season's end.

Even though the Exchange was one of the few organizations that discontinued shipment of fruit immediately following the freezes, its record for the season was extremely good. The board adhered strictly to a policy of doing everything possible to sustain the reputation of Exchange brands. On the whole, it would appear to the writer that the efforts of the Exchange in maintaining its grade were successful. There seems to be little question that SEALD SWEET held its reputation for careful efficient grade and pack in the nation's major markets.

It is interesting to note that more than 50 per cent of the Exchange's sales during the 1926-27 season were private, with about 42 per cent of the total tonnage going into fruit auctions at terminal markets. The marketing of low grade, bulk fruit, and nonstandard varieties not regularly handled by the sales department presented during this season unusually difficult problems, which resulted in the establishing of a new department within the sales department to dispose of the fruit. Although this operation posed special problems, it is apparent that the new setup was fairly satisfactory to the board.

During the 1926-27 season, the Exchange made its first major strides toward export shipments of sufficient volume to make an impression on the total Florida citrus crop. Several proposals by the Exchange for the organization of an export association representing all Florida shippers in export matters were apparently rejected by the rest of the industry. The Exchange, feeling that the sales possibilities of European markets were too great to overlook, proceeded to promote and exploit its SEALD SWEET brands on foreign markets. Nine countries in Western Europe were entered by SEALD SWEET grapefruit during this season, including markets at London, Liverpool, Birmingham, Manchester, Newcastle, Hull, Hamburg, Copenhagen, Antwerp, Paris, Glasgow, Rotterdam, Cardiff, and Oslo.

Promotionally, papers still on file from this era indicate that the Exchange conducted exceptionally successful publicity campaigns to support its export program. With the advent of the great influenza epidemic in Great Britain during this season, publicity stories featuring the value of grapefruit for "flu" were released and widely printed by European newspapers. In the words of C. C. Commander: "The effect of this work was all that could be expected, on the part of the trade as well as the consumer. It is conservative to say that $100,000 would not have purchased the advertising value of these news articles."

The 1926-27 season also produced much activity in the matter of marking the Exchange's brand, SEALD SWEET, on each piece of fruit. By the end the season, the board received a report from its general manager indicating that while there was general agreement that the marking of fruit had become a necessity, the mechanical problems of accomplishing this had not been surmounted. Some houses had received good results from a marker known as the Hale machine, but others had not been able to use this particular type. Testing was under way, however, on an Ahlberg machine which appeared to most experts to be of superior advantage.

Staff members of the Exchange had been instructed by the board early in the season to work with Commissioner of Agriculture Nathan Mayo to obtain passage of whatever laws were felt necessary for the benefit of the industry. Proposals favored by the Exchange for legislation were prevention of the use of arsenic sprays and of the shipment of frost-damaged fruit, provision of a strengthened green-fruit law, and passage of an act to standardize the grades of citrus leaving the state.

A review of citrus legislation passed during this season indicates that a law was passed by the legislature preventing the use of arsenic compounds in insecticides. A law was passed which would efficiently control the frozen fruit situation, and reconsideration of the green-fruit law led the legislature to extend the period of its coverage until December 1.

With regard to standardization of grades, General Manager C. C. Commander, in his annual report to the Board, said:

> Much as a standardization act was needed, other operators in the state were not favorable to it.
>
> A proposal to obtain a law providing certain maturity tests for tangerines was also turned down by the Fruitmen's Club. The failure of other operators to support these two phases to promote the industrial welfare of citrus, together with similar flat refusals of other important factors such as the state-wide advertising campaign, the export association, etcetera, leads this office to question seriously the value of this organization to the industry and to the state. The Fruitmen's Club has not functioned in the successful solution of any major problem which has faced the industry in the past.

Not much space has been devoted in this history to the considerable criticism of the Exchange by outside interests, or to the sometimes violent industrial wars which raged throughout the industry during the years that followed the organization of the cooperative. The increasing fury of these wars, however, necessitates consideration of their existence. In this regard, we quote from the general manager's report of the 1926-27 season:

Organization work has been continued on the basis of offering growers an efficient service to handle their product. The lurid revival campaign method, which, when the excitement dies down, leaves growers cold and possibly resentful of having been forced to sign under pressure has neither been used nor is it contemplated.

During the past season, the Exchange has had the worst competition in years. And by "worst" is meant all that the word implies. This, however, is a good omen. It is in itself evidence of the fact that the work being done by the Exchange is bringing results and depriving our competition of volume. It is the hit dog which yelps.

We have made consistent gains in nearly all territories. It is believed that, as the organization reaches a point where it is able to select members instead of proselyting them, it will have a much more stable membership.

The season has been a hard one for all concerned. Under existing conditions, growers are not satisfied, but this office believes that the morale of Exchange growers is better than that of any other operator. Much fruit has been signed by the Exchange which was not in the Exchange last season. In addition, federal officials are very friendly to cooperation, advocating and giving assistance in every way possible to get greater efficiency in cooperative organizations.

The state press has been keenly alive to the situation. Everywhere, editors evidence a grasp of the problem, and with very few exceptions, see grower cooperation as the only solution.

As a result, the year passed has shown gains in the whole field of cooperative marketing, with the greatest advances made in development of quality standards which, while less noticeable than quantity of volume, is much more sound and lasting in its effects.

In its advertising program for the 1926-27 season, the Exchange concentrated on three major objectives: the increase of consumption and demand for all citrus, the creation of a preference for Florida citrus, and stimulation of a recognition on the part of both the trade and the consumer for SEALD SWEET as Florida's finest fruit.

The program was created, directed, and placed by the Exchange's advertising department under John Moscrip and the Erwin, Wasey and Company advertising firm, one of the largest agencies of its type in the nation at that time. Magazines of both general and specific nature were used as the basic media. From the media reports of this time it would appear that general copy was placed in *Good Housekeeping, Woman's Home Companion,* and

American Magazine. To reach what must have been a fairly well-defined audience with special copy requirements, the Exchange relied on *Physical Culture Magazine. Life* and *Judge* were used to stimulate the use of SEALD SWEET citrus juice by the large, inviting cocktail and alcoholic beverage market. Total circulation of the magazines used by the Exchange during this period was nearly 6,000,000 per month. It was the policy of the Exchange to supplement and strengthen this use of magazines with more localized advertising in newspapers where it was felt that this would be to a decided advantage.

Probably one of the most important functions of the advertising department at this time was the maintainance of dealer service crews operating out of New York, Chicago, Cincinnati, Detroit, and Boston. In addition to these field teams at work in the United States, the Exchange also conducted a similar program in Great Britain. A rather complete report of the activities of dealer service crews during the 1926-27 season indicates that more than 5,926 retailers and 243 jobbers were interviewed during the period. Two hundred and twenty-one SEALD SWEET window displays were installed in addition to the distribution of 58,179 individual pieces of SEALD SWEET display material. Seventy-six markets in the United States and twelve in Europe were entered by Exchange dealer service men during the season.

The following extract from a report by General Manager C. C. Commander to the board of directors in May, 1927, cogently characterizes the advertising program for the 1926-27 season:

> The advertising activities of the Exchange have been effective as far as they have gone. It is readily evident, however, that to do the job for the industry which advertising alone can do, far more advertising than the Exchange can now support is necessary. One method of obtaining this increase would be by means of a statewide citrus campaign supported by a tax on at least 75 per cent of the fruit. This can be obtained at present only through the cooperation of other operators. This cooperation, as is evidenced by the action of a year ago, is not available.

Other promotional undertakings by the Exchange during the season included a premium program involving a new type of juice extractor which has been mentioned in earlier chapters. By the close of the 1926-27 season, the extractors were in the homes of

16,324 consumers. A financial report with regard to the extractors indicated that the premium department, which had started with an outlay of the year prior of $25,000, had cut its deficit from $34,900 to less than $4,170 by the close of this season.

A report of the premium promotion at the close of the season indicated that virtually all extractors placed had gone into private homes, although a few had been placed for fountain use. Nearly 25 per cent of the total placements of the extractors had gone into citrus-producing areas. Continuing, the report said in part that "distribution principally is this country, yet it can be safely said that the sun always shines on the SEALD SWEET extractor. It is in use in every state and territory of the United States. Extractors will be found in England, Ireland, Scotland, France, Italy, Spain, and other European countries. Japan recently began their use. Practically every country in the South and Central Americas has some of them in operation. Remote, frost-bitten northern Alaska and Cape Town, the southern extreme of Africa, enjoy its use. In effect, the extractor is functioning as an advance agent of the SEALD SWEET brand. Each extractor bears the stamp SEALD SWEET and in this is a medium of introduction and a continuous silent salesman for our brand."

The Exchange's traffic department also had an active year during the 1926-27 period. From September 1, 1926, to May 1, 1927, a total of 2,046 claims amounting to $55,075.90 were collected for Exchange growers by the traffic department. The Exchange participated in the operations of the Growers and Shippers League during this season. E. D. Dow, traffic manager, had been for some time a member of the executive committee of the League, and records indicate that the Exchange had contributed a large share of the funds with which the work of the League was continued.

A note from the production department indicates concern in Florida regarding the growing Texas grapefruit production. A table apparently produced by the production department showed that Texas was steadily making headway and was soon to become an important grapefruit producer. The table indicates that in the grapefruit-producing areas, including Florida, Texas, California, Puerto Rico, and Arizona, a non-bearing acreage amounted to nearly one-half the total citrus acreage. This fact was felt to be of extreme importance to Florida, for it was seen as an indication of the increasing competition for grapefruit markets.

The Growers Loan and Guaranty Company, which had by now been in operation for several years, lent well over $1,000,000 to Exchange growers during the 1926-27 season, and made what it felt to be satisfactory connections with Northern banks with which the Florida Citrus Exchange did business at the time.

A brief report of the Exchange Supply Company at the close of the season had this to say: "The Exchange Supply Company, also a subsidiary of the Florida Citrus Exchange, functions as a purchasing agent for associations and packing houses. It has continued its liquidation of holdings and is in excellent shape."

With regard to the canning of grapefruit, it will be remembered that the Exchange had led the way in the establishment of canning facilities for this purpose. This is what the situation looked like at the close of the 1926-27 season, according to a paper on record with reference to this matter:

The canning of grapefruit has developed into an industry of considerable importance. Not only is it becoming vital to the fresh fruit sale of citrus in that it removes from the box-lot market off-size, off-color fruit which would tend to bring down the prices of fresh fruit, but it is also obtaining for Florida grapefruit a distribution which could not otherwise be obtained. It is in this latter respect an excellent forerunner for the stimulation of demand in the creation of new consumers.

Four canneries operated during the season 1923-24 to use 98,986 field boxes of grapefruit. During the season just completed (1926-27) thirteen canneries used 533,826 boxes of grapefruit—over 500 per cent increase in four seasons. Five new canneries are now under construction and, it is understood, will be ready for operation next season.

If the quality of the product is maintained by the executives directly concerned with the production of canned grapefruit, a very bright future is in store for this industry. It is not unreasonable to assume that within a comparatively few seasons, the canning industry will be developed to such an extent that it will take from 25 to 33 per cent of the Florida grapefruit crop.

The canneries paid about 50 to 55 cents per box for culls and from 70 to 85 cents per box for choice fruit at the packing house. Large sizes were in demand.

CHAPTER 14

1927-1928

THE ANNUAL meeting of the board of directors, Florida Citrus Exchange, for the 1927-28 season was held on June 7, 1927.

Credentials of individual members of the board were read, and board members were seated as follows: W. O. Talbott of Dade County Citrus Sub-Exchange, A. J. Dozier of DeSoto County Citrus Sub-Exchange, B. F. Stewart of Charlotte Citrus Sub-Exchange, L. M. Hammel of Hardee Citrus Sub-Exchange, F. C. W. Kramer, Jr., of Highland Citrus Sub-Exchange, W. J. Ellsworth of Hillsborough Citrus Sub-Exchange, H. H. Kellerman of Indian River Citrus Sub-Exchange, Josiah Varn of Manatee County Citrus Sub-Exchange, J. C. Merrill of Marion County Citrus Sub-Exchange, F. A. Rundle of Orange Citrus Sub-Exchange, W. W. Yothers of Lake Apopka Citrus Sub-Exchange, H. C. Tilden of Seminole-Orange Citrus Sub-Exchange, John S. Taylor of Pinellas Citrus Sub-Exchange, Vet L. Brown of Polk County Citrus Sub-Exchange, John A. Snively of Florence Villa Citrus Sub-Exchange, R. O. Philpot of Lake Region Citrus Sub-Exchange, Jay Burns, Jr., of Ridge Citrus Sub-Exchange, C. H. Walker of Scenic Citrus Sub-Exchange, H. E. Cornell of the Winter Haven Citrus Sub-Exchange, and R. J. Kepler, Jr., of St. Johns River Citrus Sub-Exchange. H. G. Putnam was subsequently seated from Indian River Citrus Sub-Exchange.

Special directors for the 1927-28 season were W. E. Lee, F. S. Ruth, W. J. Howey, D. C. Gillett, and M. G. Campbell.

At the organizational meeting of the new board, E. L. Wirt was re-elected to the presidency and as chairman of the board, O. M. Felix was elected as secretary, F. C. W. Kramer as first vice-president, John A. Snively as second vice-president, H. C. Tilden as third vice-president, and C. H. Walker as fourth vice-president.

Pursuing a course that was to be followed for many years, C. C. Commander was re-elected by the board to serve as general manager. Other staff re-elections included George A. Scott as sales manager, A. H. Blanding as production manager, E. D. Dow as traffic manager, John Moscrip as advertising manager, and W. T. Covode as cashier.

One interesting action at this meeting was the reading of a resolution prepared and submitted to the Exchange by the Bradenton Citrus Growers Association. The resolution was as follows:

WHEREAS it is common knowledge that it has been the custom of many growers in the past to use the FLORIDA CITRUS EXCHANGE as a tool in their dealings with speculators negotiating for their crops, and

WHEREAS the said growers knowing full well that they could avail themselves of the privilege of membership in the FLORIDA CITRUS EXCHANGE at any time, have withheld joining the FLORIDA CITRUS EXCHANGE until all negotiations with speculators have failed, or, said growers were successful in disposing of their fruit to the said speculators, and,

WHEREAS the FLORIDA CITRUS EXCHANGE has played the go-between and has been the goat long enough, therefore,

BE IT RESOLVED that the BRADENTON CITRUS GROWERS ASSOCIATION put itself on record as condemning such practices and will not receive new members after September 1st, of each year, except owners or agents of newly acquired groves, provided this can be made a statewide movement, and we urge the FLORIDA CITRUS EXCHANGE to use every effort to make this a statewide movement.

Done by the BRADENTON CITRUS GROWERS ASSOCIATION at their annual meeting held on May 3, 1927.

(Signatures) W. J. Sanborn
A. O. Kirkhuff

The resolution was not acted upon at this meeting, nor is there indication in the records of later meetings that it was acted upon in any way.

In September, 1927, the Exchange entered into discussions with a firm known as Liquid Dehydration Corporation of Chicago, with regard to the dehydration of citrus fruits into a powder. While some disagreement existed between the Exchange and the dehydration firm concerning the scope of research and the responsibilities of the Exchange, records indicate that the Exchange finally agreed to provide fruit for this purpose and to conduct tests within the industry on the potential of the finished powdered product.

The advertising budget for the 1927-28 season was placed at $208,000, and included expenditures for trade-paper, magazine, and newspaper media advertisements as well as for display materials, dealer service activities, and a Florida advertisement program designed to obtain a favorable press and to secure additional tonnage for the Exchange.

The Exchange also ventured into the preservative dip process in an effort to extend the shelf-life of fresh citrus. Under an arrangement finally extended to the Brogdex Company, Exchange packinghouses began to use a borax solution dip under a per-box charge for the use of that firm's patented process.

It has been mentioned in preceding pages that the Exchange, through its rapid development into what was the most important single element of the industry at this time, was on many occasions in direct conflict with other factions. The 1927-28 season was a period of such conflict, triggered generally by the efforts of the industry to establish a citrus clearing house through which the control of distribution by volume shipments and market allocations, as well as a standardization of grade and pack, could be accomplished, and a general commodity advertising campaign for all Florida citrus could be developed.

A verbatim report of Exchange policy and feeling by General Manager C. C. Commander presents clearly some of the difficulties of these times. His report is as follows:

There are other factors of development which are important, but dovetail into these fundamentals. Of these lesser points, control of prices and development of foreign markets are perhaps the most important.

In its work to bring about these conditions in the Florida citrus industry which would react in favor of the citrus grower, the Florida Citrus Exchange has long recognized these fundamentals as the first steps necessary. It has adopted and continually used them in its own organization. To be effective, however, such principles of control must be applied to at least 75 per cent of the total volume rather than to the 39 per cent which is controlled by the Exchange.

The adoption of such a merchandising program for that volume controlled by other operators in the industry has presented an almost insurmountable task. Many plans having this end in view have been formulated and variously supported by those both within and without the industry. Some of these have had their origin in purely selfish business reasons. Others have been sincerely designed to be of benefit.

Perhaps the first of these plans was what is now known as the Jardine Clearing House plan. It originated outside of immediate citrus circles and obtained definite headway when Mr. Mayo, Florida Commissioner of Agriculture, and Mr. Jardine, United States Commissioner of Agriculture, called a meeting of all operators in the industry in Washington on June 13, 1927. In this meeting, Mr. Jardine discussed his ideas of a plan which would solve many of the difficulties facing the Florida citrus industry. Mr. Tenney, a subordinate of Mr. Jardine, gave the latter's ideas a definite form, which he later presented to a meeting of the Fruitmen's Club in Florida.

Independent operators having failed to obtain recognition for their plan, it became necessary for the Exchange to lay the facts before the public as to just what had happened with regard to the Jardine Clearing House plan. Such action became necessary to counteract propaganda which sought to lay the blame for failure of the Jardine proposal upon the Florida Citrus Exchange. This was done in a full-page advertisement circulated throughout the citrus belt.

Thus the first plan to coordinate shipping interests on a common platform for the good of the industry ended in failure in January. Independent operators refused to join an organization which would regulate their actions for the common good of the industry and which provided penalties for non-performance.

While the Jardine plan was still alive, awaiting endorsement of the operators in the industry, the bankers of the state met and discussed the possibility of the development of some plan whereby the united action of the banks of the state might bring about those changes in the marketing machinery which were necessary. This action met with considerable approval through the state press, and for a time it was hoped that some accomplishment might be made. It was finally decided by them, however, that such progress was being made on the Jardine plan by those actively engaged in the industry that any work on the part of the bankers was unnecessary. The bankers, as a group, thus made no progress with the situation.

Following the failure of the Jardine plan and the refusal of the Florida Citrus Exchange to endorse the dummy clearing house, the committee of the Florida Citrus Exchange continued its negotiations with the Fruitmen's Club with instructions to work with independent operators to form a true clearing house along the lines originally proposed and endorsed by the United States Government, or to form a trade association which would function as such. The latter was finally accomplished and the Florida Citrus Trade Association, Inc., was formed. This organization was to be purely informative and had no powers whatever other than advisory. Although not functioning, it is at this time in existence.

About the middle of January, the Florida State Chamber of Commerce appointed a committee to investigate the citrus situation in detail. It was to formulate a plan offering the solution to the difficulties they found and to submit this plan for approval and adoption to growers and operators in the state. This committee was composed of Walter F. Coachman, Sr., of Lake Placid and Jacksonville, chairman; Dr. Burdette G. Lewis of Penney Farms; Edward W. Lane, Jacksonville, and J. A. Griffin, bankers; Joe H. Gill, Miami, vice-president of the Florida Power and Light Company, and Dr. H. Harold Hume of Glen St. Mary, a horticultural expert.

Meetings were held by this committee in Orlando, Tampa, Jacksonville, Palm Beach, and other cities in the state at which the views of all leaders in the industry were heard. After gathering this information over a period of four to six weeks, the committee then announced that it would make its report at an early date.

Before this report was made, several growers in Polk County conceived the idea of taking the burden of obtaining support for the original merchandising fundamentals provided for in the Jardine plan into their own hands. They formed at Winter Haven on February 14th what has been called the Committee of Fifty. Judge Allen E. Walker was elected temporary president with a board of directors also temporarily in office to assist him in the organization plans. After considerable discussion, this committee decided that a clearing house set up for the same purposes provided for in the original Jardine plan along practically the same lines of organization was desirable and practical. The charter, bylaws, growers' and shippers' contracts of the organization were prepared by attorneys of the Federal Department of Agriculture before the committee started out to get the necessary grower and operator support.

The committee obtained the services of Mr. Merton L. Corey and began an intensive organization campaign that is still under way.

The Florida Citrus Exchange was asked to endorse this plan. After carefully considering the charter, bylaws, and contracts of the Clearing House organization proposed by the Committee of Fifty, the Executive Committee on April 24th committed the Florida Citrus Exchange as a body to the support of this Clearing House on condition that 30 per cent of the fruit of the state other than that controlled by the Florida Citrus Exchange would be signed with the organization by July 10, 1928.

It was later deemed advisable for the committee to sign Exchange growers individually rather than have them brought into the Clearing House organization as a unit. In order that this work of the committee might progress smoothly and without interruption, the various legal and technical difficulties inherent in this plan were swept away by the Executive Committee in its meeting on Wednesday, May 9th.

At this meeting the Exchange Committee decided to sign a shipper's contract with the Committee of Fifty organization. This action would permit the Exchange to handle not only the fruit of its growers who were signed individually, but also those who did not sign. Thus, in either event, our growers would be protected.

This matter was placed in a sub-committee's hands for execution and the shipper's contract was formally signed May 12th.

The Exchange thus announced its position. It was entirely consistent with its constant endeavor to bring into action a plan whereby the merchandising fundamentals originally outlined might function for the benefit of all citrus growers.

Thus, the Florida Citrus Exchange had committed itself to the support of the Clearing House idea during the 1927-28 season. In retrospect, today's citrus industrialist can find that the Clearing House idea is based perhaps too much on the presumption that Florida growers and operators would cooperate to the degree required for success in such an undertaking.

All available information on the subject of the Clearing House plan seems to indicate that the Exchange exerted every effort to assure its success. One factor is apparent. If the required grower support for the Clearing House was obtained, it would become necessary for independent operators to join the movement. This would then provide the industry some possibility of effective operation along the merchandising lines so much sought by the Exchange at that time.

In other respects the Exchange could also find great advantage in the Clearing House plan. Relieved of its expensive commodity advertising campaign which had been conducted for so many years, the Exchange would then be in position to spend more of its funds on direct brand advertising and on brand promotions.

It is conceivable, too, that the Exchange felt at the time that if the Clearing House idea failed, enough sympathy would have been created for the idea that the possibility of expanding its own facilities into a Clearing House sort of organization would have been extremely good.

As for the 1927-28 season itself, there is every reason to believe that the Exchange looked upon it as a good year. Returns were satisfactory and the cooperative secured favorable prices for its growers. Although some areas of the citrus belt were struck by extremely cold weather, and although some frozen fruit was again shipped to the market, considerable optimism is apparent in various publications recorded at that time.

Competitive conditions apparently remained about the same as for the season before. Nonbearing acreage of Texas, Arizona, and Puerto Rico had increased and, in the words of Commander, were "one season nearer to the addition of its volume to the American market."

During the 1927-28 season, the Exchange placed 58.2 per cent of its fruit in private sales channels and sold slightly over 40 per per cent at auction. Generally, Exchange growers received prices higher than growers shipping through independent handlers, as is apparent from a record of auction prices realized during the season.

One board member considered tactics of Exchange competitors to be somewhat less than fair during the season. "Competition, especially in private sale markets," he said, "has been not only keen but difficult because of the tactics employed by some of our competitors. In spite of the fact that it has been a sellers' market and volume considerably short of supply, many of our competitors continued their old practice of selling below sales quotations and wire, as usual, 'If Exchange sells for blank dollars, offer our fruit 25 cents less.' "

In advertising, the Exchange was forced to cancel a considerable percentage of its original program scheduled for this season because of freezes. Dealer sales work in Europe was continued with fifteen European markets entered into and thoroughly worked by the Exchange representative in Europe.

A letter on file in the official record of minutes, dated December 17, 1927, with regard to Traffic Manager E. D. Dow, is suggestive of the prestige enjoyed within the industry by the Florida Citrus Exchange. Written by Frank Kay Anderson of the *Citrus Industry Magazine*, the letter was directed to the general manager and asked that Traffic Manager Dow be permitted to submit himself as a candidate for election to the Florida Railroad Commission. "In addition to his undoubted qualifications for the post, Mr. Dow is possessed of a record and certain personal attributes which would be substantial assets to him as a candidate. Among our Florida perishable traffic men, he is the one logical candidate whose election should be most easily obtained," wrote Anderson.

A subsequent entry in the minutes placed the Exchange on record as declining the request, an action obviously designed to establish a precedent in matters of this nature.

The 1927-28 season was completed with the annual meeting on June 7, 1928. Directors and their sub-exchanges seated at this meeting were John A. Snively of Florence Villa, B. F. Stewart of Charlotte, W. O. Talbott of Dade, A. J. Dozier of DeSoto, L. M. Hammel of Hardee, F. C. W. Kramer, Jr., of Highland, W. J. Ellsworth of Hillsborough, Homer Needles of Indian River, W. W. Yothers of Lake Apopka, R. O. Philpot of Lake Region, Josiah Varn of Manatee, Walter R. Lee of Marion, F. A. Rundle of Orange, Vet L. Brown of Polk, Jay Burns, Jr., of Ridge, C. H. Walker of Scenic, H. C. Tilden of Seminole-Orange, J. E. Bartlett of St. Johns River, H. E. Cornell of Winter Haven, and John S. Taylor of Pinellas. W. E. Lee and F. S. Ruth were seated as special directors.

E. L. Wirt was again elected as president and chairman of the board, O. M. Felix as secretary, F. C. W. Kramer as first vice-president, John A. Snively as second vice-president, H. C. Tilden as third vice-president, and C. H. Walker as fourth vice-president. C. C. Commander was re-elected as general manager, George A. Scott returned as sales manager, A. H. Blanding as production manager, E. D. Dow as traffic manager, John Moscrip as advertising manager, and W. T. Covode as cashier. Judge William Hunter was subsequently reappointed as legal counsel for the Exchange.

Left, the Citrus Exchange Building in 1922 was the very latest in modern office buildings. It was to house the Exchange for many years before the organization's growth required additional quarters. In the heart of downtown Tampa, the building is now occupied by Maas Brothers, Inc.

The Packing House, above, was typical of Exchange houses at the beginning of the 1929-30 season.

The old Eagle Lake Packing House, below, was converted into Florida's first processing plant for grapefruit sections. First pack of sections was made during the 1921-22 season; however, the photo predates the conversion of the house to a processor facility.

One of the many guard posts established in 1929 to prevent movement of citrus out of ar
infested by the Mediterranean Fruit Fly.

Governor John W. Martin, staff and party, are shown as they visited the Winter Haven Citr
Growers Association in 1927.

CHAPTER 15

1928-1929

IN SPITE OF the freezes and storms that had periodically plagued the Florida citrus industry through the years, the 1928-29 season was more memorable because of adversities than any season on record until that time.

On Saturday, April 6, 1929, J. C. Goodwin, nursery inspector of the State Plant Board, noticed what he thought to be unusual spots in grapefruit which came from the grove of the United States Department of Agriculture Experiment Station. Upon examination he found maggots which were different in appearance from those normally found in decayed fruit. He immediately forwarded a specimen to Washington for positive identification.

Washington's reply was prompt. It was the larvae of the Mediterranean fruit fly. The Exchange, along with every other handler, grower, and shipper in Florida immediately became involved with an enemy possessed of the capability of wrecking the entire Florida citrus industry. Eradication forces were organized under the direction of Dr. Wilmon Newell, plant commissioner. Quarantine regulations were drafted and placed in operation by the United States Department of Agriculture, and State Plant Board inspectors trooped through every grove in the state to search out additional infestations. Infested fruit was destroyed by the ton. By May 20 a so-called nine-mile-limit area quarantine had been placed in effect surrounding each infested area of the citrus belt. Guards and inspectors had established effective posts to prevent the movement of citrus out of the infested areas. Meantime, the federal government rushed a legislative measure through both houses of Congress placing $4,250,000 at the disposal of state authorities. Florida appropriated another $500,000 and offered more if the need arose.

As states in the South and West closed their borders to Florida citrus in fear of the spread of the fruit fly into their large agriculture areas, economic consequences were apparent. In the expectation of a more general quarantine on all Florida citrus, scores of shippers began glutting the markets with their fruit without regard to market capacity.

As this situation intensified, General Manager C. C. Commander warned Exchange members of the drastic break in prices which was sure to follow such a glut. The Exchange, acting on the spur of the moment, promptly secured cold-storage facilities for a large part of its tonnage and simply "rode out the break in prices." The trend toward lower prices reversed sharply as additional quarantine action was placed in operation and, as quickly as the glut had occurred, a sudden shortage saw the Exchange releasing much of its stored fruit at favorable prices. Reporting later to the board on this matter, Commander said, "This plan, developed and placed in operation practically overnight, was more than moderately successful. It is one which is worthy of the careful consideration of other operators. Coordinated action on this plan would save many market depressions caused by excessive volumes during any average season."

From this viewpoint, while not denying the seriousness of the Mediterranean fruit fly, various Exchange officials pointed out that some favorable aspects were resulting from the situation. Certainly groves would be more carefully inspected and cared for, and licensing of packinghouses, it was felt, would result in greater standardization and control. Finally, the shipments of bulk fruit which had caused serious handicaps in the profitable sale of boxed citrus had been eliminated under the quarantine regulations.

The drastic embargoes placed on shipment of Florida citrus into other states would, in all probability, serve to force a much more rapid development of methods of distributing the fruit parts or its juices in a prepared and more nonperishable form. By this time, grapefruit sections had been successfully packed and distributed for a number of years with ever-increasing volume. Even canned grapefruit juice was gaining in popularity. There was also a form of orange concentrate being sold for use by soda fountains, bakeries, and other institutional use, but, as one Exchange official remarked at the time "the dissimilarity between the fresh fruit and this concentrated product greatly limits its commercial importance."

Thus, the Mediterranean fruit fly played its role in the development of the Florida citrus industry by posing problems that were to be answered only by research and unselfish industrial progress. Distribution of preserved citrus products was a natural solution to the embargo difficulties of that era, and the Exchange commenced work immediately on several experiments of this nature.

In addition to the fruit fly, several other adversities combined to make the 1928-29 season one of the most difficult the industry had ever faced. Prior to the discovery of the fly, the greatest single factor taxing marketing agencies at the time was the exceptionally large crop of citrus fruit produced and shipped to domestic markets. The combined production of citrus areas feeding American markets reached more than 60,967,330 boxes. Of this total, California produced a little more than 60 per cent, and Florida produced slightly over 30 per cent. Texas, Arizona, Alabama, Louisiana, and Puerto Rico combined to ship the remainder of the total.

The total of 60,967,330 boxes of citrus shipped to domestic markets during the season was 33.8 per cent greater than that of the 1927-28 season, and 19.1 per cent greater than the previous large production season of 1923-24, when markets consumed a total of 48,102,680 boxes.

The marketing situation was also apparently further involved in adversity for Florida because of the relatively poor quality of the season's production. Then, too, the rise of the commercial standing of Texas grapefruit was evidently beginning to be of consequence to Florida. For the first time, successful distribution of Texas fruit on a large scale was beginning to take over markets which had heretofore been exclusive for Florida grapefruit. To make matters worse for Floridians, the quality of the Texas product was in no way inferior to that of Florida producers. Of further concern was the fact that only an estimated 11 per cent of the estimated 83,500 acres of grapefruit plantings in Texas were beginning to bear.

On other fronts, records indicate that the Florida Citrus Growers Clearing House Association, mentioned in the preceding chapter, had been activated and was functioning. While its effectiveness was apparently not all that had been expected by the Exchange, it seems to have served well in pack standardization and in commodity advertising.

In sales for the 1928-29 season, the Exchange increased 61.35 per cent over the preceding season with particularly large gains in grapefruit and tangerines. Forty-nine per cent of Exchange sales for the season were placed in private sale, while 48 per cent was sold at auction.

On the export market the Exchange continued its program of development of foreign trade. This report on the progress of the export trade is contained in the annual report of the 1928-29 season:

> The SEALD SWEET trademark more firmly than ever is becoming established in the minds of the European trade and consuming public as a mark of quality grapefruit. This condition is fostered by consistent sales and display advertising work on the part of the Exchange.
>
> The European demand for SEALD SWEET grapefruit again increased. More than twice as much fruit was shipped to Europe this year than during last season. Car lot supplies have been going forward steadily to Liverpool, London, Glasgow, and Hull. Less than carload lots have been going forward regularly from New York to Copenhagen, Antwerp, Rotterdam, Hamburg, Oslo, and Newcastle.
>
> From these distributing points, SEALD SWEET fruit has entered the lesser markets of ten countries in Western Europe during the past season. The important markets of these countries—England, Scotland, Germany, Belgium, Sweden, Norway, Holland, Denmark, France, and Finland—have received the greater proportion of the export volume.
>
> The importance of the United Kingdom as a market outlet is shown in distribution records which reflect that in 1923, the year of the Exchange's first venture into export trade, a total of 10,000 boxes of Florida citrus was placed on that market. During the 1928-29 season, the United Kingdom utilized well over 200,000 boxes.

The 1928-29 season saw practically no change in the physical composition of the field organization of the Florida Citrus Exchange. At the season's end, it was composed of eleven active sub-exchanges comprising eighty associations and eight special shippers, operating seventy-six packinghouses. Twenty-six of the Exchange houses had by this time installed precooling plants, and fifty of them were operating marking machines for stamping the SEALD SWEET name on individual fruit.

In advertising, the Exchange adopted a "one-fourth more juice" slogan after conducting research into the chemical analyses of Florida and California orange juice, the results of which indicated a definite average superiority in both quantity and quality for Florida oranges. This was the basis of the Exchange's over-all advertising and merchandising effort for the 1928-29 season. As in the past, the Exchange relied almost exclusively on newspaper and magazine advertising, although it ventured cautiously into the relatively new field of radio in the Denver, Colorado, area.

Fourteen dealer service crews composed of thirty-four regular personnel and additional local help were maintained during the 1928-29 season by the Florida Citrus Exchange. Two of the crews functioned in the mid-western division out of Chicago, two out of Detroit, two out of Cincinnati, three in the Eastern division, one in the New England division, and four crews in the two Southern divisions.

Combined, the merchandising force of the Exchange covered 997 jobber contacts and 34,346 retailers. They placed Exchange materials in 10,785 store windows and utilized a total of 180,977 pieces of advertising display material.

By the conclusion of the season, four gigantic fruit and vegetable terminals were nearing completion at Detroit, Cleveland, Pittsburgh, and Providence. The Detroit terminal was expected to open on July 1, 1929, with thirty-eight acres of terminal warehouse space available. Two two-story terminal houses were to form the major structures, with the possibility of display of three hundred cars of fruit and vegetables at a single showing.

While there can be no doubt that the 1928-29 season was complex and variable, there is reason to believe that, on the whole, cooperatives survived the season in extremely good shape. This is supported by a report of the general manager to his board in May, 1929:

From a financial standpoint, packinghouse operations have been in general the finest in the history of the organization. Most houses, because of the volume handled, have been able to keep their costs at a very satisfactory figure. This will enable them to make substantial refunds to their growers. The Exchange organization as a whole is more closely united than ever before. It is in excellent position to present a united front to the many serious problems which present themselves for solution during the coming season.

Certainly not the least of these problems was the uncertainty of the effect of the fruit fly on industry. The crop outlook was enigmatic, for it was impossible at the close of the 1928-29 season to estimate the following season's marketing potential on the basis of expected production alone. Into every estimate of available fruit came the unanswerable question of the effect of the fly on Florida's production. Of the known elements affecting the crop outlook for the following season, however, it is interesting to note that the set of fruit was much lighter than the year preceding, although the Valencia outlook indicated a fairly good crop. In some sections grapefruit bloom was reasonably good, while in others it was reported as very short.

These conditions were prevalent as the board of directors completed their work at the conclusion of the 1928-29 season. The Florida Citrus Exchange, on its twentieth anniversary, remained the stronghold of the cooperative movement in Florida citrus. Its successes were many and its infrequent shortcomings had been largely honest miscalculations of man and nature. Ahead somewhere still veiled in illusiveness, was the answer to the aging problem of volume control so persistently sought by the Exchange during its formative years.

CHAPTER 16

1929-1930

THE ANNUAL meeting of the board of directors at the close of the 1928-29 season was held on June 4, 1929. Directors seated at this meeting were J. O. Carr of Charlotte Citrus Sub-Exchange, W. O. Talbott of Dade County Citrus Sub-Exchange, A. J. Dozier of DeSoto Citrus Sub-Exchange, W. J. Ellsworth of Hillsborough Citrus Sub-Exchange, Homer Needles of Indian River Citrus Sub-Exchange, W. W. Yothers of Lake Apopka Citrus Sub-Exchange, F. C. W. Kramer, Jr., of Lake County Citrus Sub-Exchange, F. P. Goodman of Lake Region Citrus Sub-Exchange, R. K. Thompson of Manatee Citrus Sub-Exchange, W. R. Lee of Marion Citrus Sub-Exchange, F. A. Rundle of Orange County Citrus Sub-Exchange, J. S. Taylor of Pinellas County Citrus Sub-Exchange, Vet L. Brown of Polk Citrus Sub-Exchange, John D. Clark of Ridge Citrus Sub-Exchange, C. H. Walker of Scenic Citrus Sub-Exchange, A. W. Hurley of Seminole-Orange Citrus Sub-Exchange, J. E. Bartlett of St. Johns River Citrus Sub-Exchange, and H. E. Cornell of Winter Haven Citrus Sub-Exchange.

Special directors seated were F. S. Ruth, Jay Burns, Jr., W. J. Howey, and W. E. Lee, and Clinton Bolick was seated as an associate director.

Election of officers for the 1929-30 season was held on July 5, 1929. E. L. Wirt was returned to the top office as president and chairman of the board, John A. Snively was elected as first vice-president; John S. Taylor as second vice-president; W. W. Yothers as third vice-president; and C. H. Walker as fourth vice-president. C. C. Commander was returned as general manager, O. M. Felix as secretary, George A. Scott as sales manager, A. H. Blanding as production manager; John Moscrip as advertising manager; E. D. Dow as traffic manager, and W. T. Covode as cashier. Judge William Hunter was again re-elected as legal counsel for the Exchange.

The 1929-30 season according to all available records was affected by three unusual but major factors. The quarantine regulations placed on the shipment of Florida citrus fruit because of the Mediterranean fruit fly infestations were without doubt the most serious difficulty of this period. In addition to this, records indicate that quality was again generally below standard levels with respect to both appearance and eating quality. The third factor, however, apparently tended to offset the first two disadvantages. The crop volume for this season was one of the lightest in ten years. Comments and papers from recognized leaders in the industry at this time seem to verify the assertion that without this saving feature it would have been almost impossible to have marketed the entire crop at a profit to growers.

Not only was Florida's production low, but the total volume of citrus produced by all citrus areas supplying American markets was considerably reduced from the 1928-29 season. Statistics from this period indicate that total production of California, Florida, the Rio Grande Valley of Texas, Puerto Rico, and the Alabama-Louisiana-Mississippi production area was slightly more than 20 per cent less than the production of the previous year.

With regard to the Exchange itself during the 1929-30 season, the organization, in spite of the short production, handled 39.5 per cent of the state's total fresh-fruit sales—an increase of 8 per cent over the preceding year. To those familiar with the early development of the cooperative movement in the state, this would seem to be a reversal of the usual situation in which marketing cooperatives increased their percentage of control during heavy crop years when returns were low and the speculative element inactive. It seems apparent that this increase in tonnage was due in great part to the timely assistance of the Federal Farm Board, which, under authorization of the Agriculture Marketing Act, approved loans to the Exchange totaling $3,300,000. These loans, available at low rates over a relatively long period of time, made possible the financing of mergers of large independent grower-shipper interests with the Florida Citrus Exchange. Even though these mergers themselves were to effect some changes in policy and operation in the Exchange organization and were to place new people on the board, in the interest of brevity this history must generalize to a great extent in the coverage of this era. Certain facts, however, played an important role not only in the contemporary structure of the Exchange, but also in their effect upon the future history of the organization.

Merger activity was entered into with the Florida Citrus Exchange during this period by Chase and Company, Florida United Growers, International Fruit Corporation, Lucerne Park Fruit Association, and others. It should be noted that much of this merger activity was apparently brought about through the attitude of the Federal Farm Board, which had decreed that the Florida citrus industry could qualify for federal loans only by some system of consolidation which would place a greater percentage of the Florida crop under a single marketing control. It must be remembered that these were the years of the Great Depression, and that the extent of the government's assistance even on the most difficult terms was considered so important as to cause the creation of interest in almost any program that would qualify the industry for this assistance. Thus, as mergers were consummated to meet the exigencies of the times, some of these transactions were to last only a few short years while others were permanent and long lasting.

One of the immediate effects of the consolidation was the redesign of certain phases of the charter and bylaws of the Florida Citrus Exchange to accommodate independent grower-shipper organizations. In this process, and through agreements with the Chase organization at the time of its merger with the Exchange, J. C. Chase of that firm became the Exchange's president, an office previously held concurrently with that of chairman of the board. E. L. Wirt, who had vacated his position as president to make way for Chase to assume that position, remained as chairman, and D. C. Gillett of Tampa was seated as a director representing International Fruit Corporation.

While the merger activities may have caused considerable cooperation among those interests basically involved, the industry was on the whole in a state of almost complete antagonism with regard to both the merger actions and the loans of the Federal Farm Board. Charges and countercharges were hurled across the citrus belt in verbal violence that was seldom equalled in the rocky course of development of the industry. For the most part, the controversy involved the Exchange and nonmembers, both cooperative and independent. But the repercussions reverberated even within the relative unity of the Exchange itself. Charges of misconduct of Exchange business by its officials were aired publicly as was the suspicion of misappropriation of Exchange funds by the board. Newspaper advertisements, editorials, industry meetings, and ominous rumors stacked one upon the other until tension dictated that these troubled waters must necessarily be calmed if the industry was to survive.

The Exchange was moved to appoint an investigation committee composed of Edward R. Bentley, Rupert Smith, R. P. Burton, F. E. Grigham, J. C. Palmer, and William Drew to investigate charges against the Exchange organization. None of the committee was a member of the board of directors, and the commitee itself was asked to operate entirely on a nonpartial basis, to hold hearings, and to take testimony. The findings of this committee, while not favorable to the Exchange in all respects, did much to heal the rift and pacify the critics.

This era cannot be passed over without, in all fairness, considering that the mergers, combined with the additional financing they afforded the Exchange, were of most important consequence to the rest of the industry. This fact, when viewed with regard to the economic depression, existing nationwide, is sufficient to reconstruct a situation which provided ideal elements for an industrial battle. It is doubtful that any of the participants emerged without long-lasting scars, and it must be noted that even though it was apparent eventual victor in the conflict the Florida Citrus Exchange has never completely resolved some of the differences that were born of this era.

Meanwhile, the Mediterranean fruit fly continued to hamper marketing operations throughout the state. However, according to a report of General Manager Commander on May 1, the last major infestation was found on January 31, 1930. The Med-fly situation was therefore most encouraging by the close of the 1929-30 citrus period, and citrus people throughout Florida were cautiously optimistic about the future.

Information available indicates that the Exchange continued to support the Clearing House with the feeling that trade matters affecting the industry as a whole were properly conducted through that organization.

One of the most important factors during this season, insofar as the utilization of fruit was concerned, was thc arrangement made by the Exchange with canners to supply grapefruit for canning purposes at stabilized prices over a five-year period. Under this arrangement, made initially with the Floridagold Citrus Corporation at Eagle Lake, canners agreed to purchase their raw grapefruit requirements exclusively from the Exchange. For the 1930-31 season, packinghouse f. o. b. prices for the fruit would be 90 cents per box, with this figure rising to $1.00 per box in the 1932-33 season, where it would remain for the 1933-34 and 1934-35 seasons. While there is evidence that this arrangement did not live up to long-range expectations, it is interesting to note that this occasion marks, perhaps, the first major recognition of the importance of the canning trade to the economy of the Florida citrus industry.

There was a continuance of the Exchange's major market area merchandising program under the same general plan as in the preceding years, although the difficulties of this period seem to have necessitated great care in the release of funds for this purpose.

Export shipments by the Exchange were greatly reduced during this season, an obvious result of the short crop. Although plans had been made at the close of the 1928-29 season to ship a considerable volume abroad, it appears to have been apparent early in the new season that these plans would of necessity be shelved. In addition to other factors which precluded volume exports, Florida grapefruit ran heavily to large sizes during the 1929-30 season, while European requirements called for heavy commitments of small-sized fruit. Under a trial export plan to Buenos Aires, however, the Exchange developed the introduction of Florida grapefruit into that country. The plan worked most successfully on a small shipment basis out of the New York port, and a survey of the results indicates that the season's operation netted a profit to Exchange growers of nearly $1.00 per box.

At the close of the 1929-30 season the Florida Citrus Exchange was composed of thirteen sub-exchanges, and a hundred and six associations and special shippers. A comparison of the organization with the previous year indicates a gain of two sub-exchanges and eighteen assocations and shippers. The over-all Exchange organization during the 1929-30 season embraced 5,964 grower-members, according to statistics compiled by the auditing department at that time. Exclusive of the acreage controlled by Chase and Company and the International Fruit Corporation, the Exchange had increased its acreage control by 12,930 acres.

It is probable that more than ever before the question of adequate financing became the most important element in the decision of growers to consider membership in the Exchange. With abnormal seasons for several prior years, growers needed desperately to arrange for financial assistance. Through the activtiy of the Growers Loan and Guaranty Company, the grower financing problem was being met with considerable success during this time. While the Guaranty Company was apparently unable to renew its lines of credit with a few of the Northern banks, it retained the majority of its connections and was successful in obtaining considerable financial assistance. With the assistance of these banks and the Federal Intermediate Credit Bank as well as the Federal Farm Board, the company was able to continue to provide aid to growers and shippers in the Exchange system.

There is little reason to doubt that the operation of the Growers Loan and Guaranty Company was particularly responsible for a great portion of the development of cooperative marketing in the Florida citrus industry at this time. The growth of the industry seems to have been directly reflected in the percentage of the total crop under cooperative control, and, during these troubled times, the Exchange's finance subsidiary served as a major inducement to growers and shippers to become associated with the large marketing cooperative.

The work of another subsidiary, the Exchange Supply Company, should also be noted in this chapter. A report of the activity of the supply company for this period indicates that it handled practically everything connected with the operation of a packinghouse. Although operated on a close margin of profit, the supply cooperative gave liberal discounts to members, and wound up the season with a small net profit. At this time, the Exchange Supply Company was in the process of liquidating its fixed holdings acquired in connection with various manufacturing departments utilized by the Exchange in prior years but discontinued as the industry continued to develop.

To summarize the status of development of the Exchange in the 1929-30 season, it might be well to review some of the major items covered in a report of a special organization committee appointed by the board. This report was rendered on December 19, 1930.

The committee does not overlook the fact that the present careful maintenance of grading standards, fair sales contacts and brand advertising have obtained for the Florida Citrus Exchange brands a considerable and enviable good will among the trade and consumers throughout the country. This good will factor is an added value in making practical the handling of a control volume by the Florida Citrus Exchange.

In addition to these well-organized facilities, the Florida Citrus Exchange has a financial subsidiary which is among the leaders of its type of organizations in the country. It commands the good will and respect of credit contacts everywhere. Its operation since its organization (1918) has been conducted in a consistent, safe manner for the benefit of Exchange grower members.

The committee has yet to find one factor advanced as a cause of the conditions existing in the industry which could not be corrected by the vestment of a minimum of 75 per cent control in the Florida Citrus Exchange. Given that control, the Florida Citrus Exchange could operate satisfactorily the basic merchandising fundamentals necessary for the sale of each succeeding crop at a profit to the producer.

It could regulate shipments, both geographically and periodically, in accordance with definitely recognizable demand. It could standardize grades and packs. It could put into operation and maintain proper methods of price quotations. It could control prices and maintain them at a level consistent with the market value of the fruit. With its solidified grower support, it could obtain enactment and enforcement of proper green fruit legislation.

All of these factors, however, depend upon the concentration of a minimum of 75 per cent control of the fruit volume year after year in the organization. The accomplishments attributed by this committee to control are not fantastical theory. The efficiency of that control is adequately illustrated in the California lemon situation. Further evidence of the practical nature of this emphasis on control is available from an analysis of the returns made by the California Citrus Growers Exchange to its growers.

California admits that it costs them more to produce citrus than it does Florida. Yet, by adequate control, their growers are paid a price consistently which nets a profit to the grower over cost of production. Florida uses the same methods of merchandising, but without the same percentage of control. It repeatedly pays the penalty.

From the foregoing it is obvious that, even in its finest hours, the Florida Citrus Exchange looked longingly to the future for a method of industrial cooperation that would calm the highly competitive wildness of the industry to the benefit of its scores of individual growers.

CHAPTER 17

1930-1931

AMERICAN citrus-producing sections raised bumper crops of citrus during the 1930-31 season, and their combined volumes were the greatest produced in the history of the industry. In Florida, where the increase was most pronounced, the rate of increase of production was 88 per cent over the previous season. California increased its production by 52 per cent, and Cuba was up an estimated 44 per cent. Texas and Puerto Rico, however, showed an average decrease of about 50 per cent.

The merchandising of this tremendous crop, which in Florida alone totaled 19,200,000 boxes of oranges and 15,800,000 boxes of grapefruit, was complicated by the continued economic depression which existed throughout the country. The purchasing power of all markets, but most particularly those primarily dependent upon the payrolls of manufacturing industries, was sharply reduced. On the other hand, according to available records, Florida's citrus crop was of unusually good quality throughout. Within the Exchange organization itself, more than 40 per cent of its fruit graded U.S. #1 quality, and 40 per cent of the remaining fruit met specifications for U.S. #2 grade.

The Mediterranean fruit fly had ceased to be a factor in the movement of Florida citrus by this time, and indications are that the entire crop of 1930-31 was moved without a recurrence of infestation. Decay factors for the season were at a minimum and losses through decay were at their lowest point in recent seasons.

The low decay percentage made possible the merchandising of fruit by various methods to the advantage of growers throughout the state. Sale of bulk fruit had, of course, been revived, as well as considerable sales in consumer bag packages.

For 1930-31 the Board of Directors for the Florida Citrus Exchange, representing the Citrus Sub-Exchanges, included: J. C. Chase of Chase, J. O. Carr of Charlotte, W. O. Talbott of Dade, Rupert Smith of DeSoto, John A. Snively of Florence Villa, R. P. Burton of Lake, Marvin H. Walter of Hillsborough, Homer Needles of Indian River, D. C. Gillett of International Fruit, J. G. Grossenbacher of Lake Apopka, R. O. Philpot of Lake Region, R. K. Thompson of Manatee, Walter R. Lee of Marion, C. A. Garrett of Orange, John S. Taylor of Pinellas, Vet L. Brown of Polk, J. D. Clark of Ridge, C. H. Walker of Scenic, A. W. Hurley of Seminole, R. J. Kepler, Jr., of St. Johns River, and H. E. Cornell of Winter Haven. Special directors were F. S. Ruth, D. A. Hunt, E. L. Wirt, L. R. Skinner, and Clinton Bolick.

Elected officials of the organization for the 1930-31 season were J. C. Chase, president; E. L. Wirt, chairman; John A. Snively, first vice-president; John S. Taylor, second vice-president; Rupert Smith, third vice-president; Homer Needles, fourth vice-president; C. C. Commander, general manager; F. W. Davis, general sales manager; J. Reed Curry, manager, Organization Department; John Moscrip, advertising manager; E. D. Dow, traffic manager, W. T. Covode, cashier; and Judge William Hunter, attorney.

Sub-Exchanges included Chase, Charlotte, Dade, DeSoto, Florence, Lake, Hillsborough, Indian River, International, Lake Apopka, Lake Region, Manatee, Marion, Orange, Pinellas, Polk, Ridge, Scenic, Seminole, St. Johns River, Winter Haven, and five additional grower-shipper organizations in the Special and Associate categories.

One of the most important actions of the Exchange with regard to the entire industry was its decision on May 15, 1930, to resign its membership in the Florida Citrus Growers Clearing House Association. This action was undoubtedly brought about because of a growing feeling within the Exchange that the Clearing House had little to offer the large cooperative in the way of marketing assistance. While the Exchange could see the direct benefit of Clearing House services within the industry, the board seems to have reached the conclusion that its marketing services failed to justify the expense. Complete withdrawal from the Clearing House was apparently decided upon only after the Exchange had presented a plan which would continue its membership for in-industry services, but which would exclude the organization from marketing services. Under this plan, the Exchange would continue its membership on a reduced assessment. The plan was not, however, acceptable to the Clearing House Association, and the Exchange withdrew its membership and severed connections with the Clearing House with full support from its associations and growers.

Thus another dream of total cooperation among the varied interests of the Florida industry was apparently dissolved into bleak reality, and the Exchange redesigned its staff and policy to fill whatever gaps were left by its withdrawal from the Clearing House.

Probably the most important development of the 1930-31 season for the Exchange was in moves to perfect its sales policies, methods, plans, and organization. Radical changes were made in personnel, one such move being the subdivision of the sales department by commodity. This change in organizational structure was designed to make possible a much closer supervision and concentration on sales of oranges and grapefruit. Division of the country into sales divisions and districts permitted an absolute localization of responsibility in the development of all possible markets for Exchange fruit.

In retrospect, it is possible today for the citrus veteran to pick flaws in this structure, and the Exchange was to find the plan unworkable before the passing of many seasons. Yet the enthusiasm with which both the board and its executives launched the reorganization is indicative of the times and the constant effort toward greater efficiency of the organization.

The realignment of the sales department resulted in the employment of F. W. Davis as general sales manager, but retained George Scott, former sales manager, as the orange sales manager and placed E. E. Patterson in the position of grapefruit sales manager. Under the revised system, weekly forecast bulletins, in addition to the daily wire service, were released by the sales department to association and sub-exchange members. These bulletins kept the entire organization fully informed at all times as to the condition of the markets and the prospective tendencies from time to time during the season. The plan was, of course, to keep association managers fully informed so as to enable them to handle their growers' crops to the best advantage. The immediate results were apparently good, for by the close of the season each division and district had seen an increase in business. While some of this was undoubtedly due to the greatly increased production of the season, there is reason to believe that the reorganization plan was at its peak efficiency in just such circumstances.

The bulk-fruit business, normally a menace to good merchandising in this period, was handled to good advantage. Bulk trade was confined almost entirely to small towns and markets where this type of shipment made possible successful competition of Florida fruit with that of California and Texas.

Through the particular methods utilized by the Exchange in handling its bulk fruit, reports indicate that practically none of the Exchange volume moved in bulk was a factor in the demoralization of fresh-fruit markets because of bulk competition. This was true despite the fact that approximately 1,200 cars of grapefruit and 1,500 cars of oranges were moved in this manner during the season.

It is interesting to note that the Exchange, during the 1930-31 season, was quite enthusiastic over its renewal of a sales contract with the Pacific Fruit Exchange. Some of this enthusiasm is apparent in a report by General Manager Commander in May of 1931:

> Final and adequate evidence of the general rounding out and strengthening of the Exchange as a sales organization is found in the renewal of the Pacific Fruit Exchange contract with the Exchange Company. That an organization of the standing and importance of the Pacific Fruit Exchange should repeatedly renew its sales contract for increasing volume is a tribute to the efficiency and ability of the sales organization owned and controlled by Exchange members.

The export volume, which had drastically declined during the preceding season, sprang to life with the high-volume season of 1930-31. Exchange export sales for the total season were three times as great as sales of any prior season on record. Export shipments were made almost wholly to London and Liverpool, with distribution from these points to Birmingham, Newcastle, Manchester, Glasgow, Paris, Hamburg, Bremen, Antwerp, Copenhagen, Rotterdam, Oslo, and Zurich. Research of the foreign trade during the season indicates that export shipments brought in almost all cases as much as the prevailing prices in domestic markets and in some instances they returned more.

The extent of the Exchange activity in the development of new methods of utilization of Florida citrus is apparent in this official annual report of the 1930-31 season:

The Exchange participation in the development of by-products industries has been of marked advantage to its growers. Benefit derived from these industries will increase as the details attendant upon the production and sale of their products are worked out and volume in their movement is obtained.

The Florida Citrus Exchange undoubtedly is responsible for the development of the frozen orange juice deal in interesting responsible well-capitalized distributing units in its manufacture and sale. Frozen orange juice would not be a commercial reality today had it not been for the investigational and research work put on this product by the Exchange over the past two years. To handle the growing volume of work attendant upon the development of this by-product, the Exchange Juice Company was organized as a subsidiary of the Florida Citrus Exchange in October of 1930.

The primary object of the organization of the company is to develop the juice end of the industry, especially frozen orange juice. Its major operations to date have been limited to research work. Investigation of the market for frozen juice indicated that a large investment in plants, advertising, etc., would be required for the preparation and sale of the product. It did not seem advisable to spend these funds if information and experience in the handling of the product could be obtained elsewhere.

The National Juice Company, a subsidiary of the National Dairy Product Corporation, had already been formed and the Florida Citrus Exchange was under contract to supply this company 250,000 boxes of juice grade fruit for its freezing operations on orange juice. The Borden Farm Products Company entered the field shortly after the National Juice Company started operations. With these well-organized and financed organizations in the field of frozen juice, it seemed advisable that the Exchange Juice Company confine its operations for the first season to research work, gaining whatever information and knowledge possible from the operations of these companies.

Considerable information has been obtained to date, but much more complete analysis of the sales value of the product will be obtained during the Summer months when there should be the greatest demand for it. Generally speaking, frozen orange juice is being favorably received in most markets, principally because of the convenience in handling. Due to economic conditions, the demand probably has not been as great as it would be were conditions normal. The low retail price of fruit this season in practically all markets has caused many prospective users of frozen juice to consider it too expensive when compared with the fresh fruit. When the Florida shipping season is over, there will undoubtedly be an advance of the retail price of fresh fruit with a consequent rise in sale of the frozen product.

Research work in this branch of the industry as conducted by the Exchange Juice Company has embraced the assembling and storing of samples from the different companies in the field to permit a comparative study of the products over a period of time. The most acceptable size and kind of container for the product has been considered. The best and most economical method of extracting the juice is another point on which investigation has been made. In addition, efforts were made to ascertain the most feasible system of freezing, storing and shipping the frozen juice. In considering work on this product it should be noted that in addition to the firms above named, several Exchange houses formed at Winter Haven the Florida Orange Juice Corporation. This plant was equipped under the direction of the Exchange engineer and froze approximately 5,000 gallons of juice before the end of the season. The quality of the product turned out at this plant was exceptionally good. At the end of the season their management had reduced the cost of operation to a point lower than that maintained in either of the operations conducted by the above-mentioned firms. This plant will be ready to start operations as early as needed during the season of 1931-32.

Research work has also been conducted on canned and bottled juices. Numerous samples of all known brands on the market were stored and are being used for comparative tests over definite periods of time.

Similar work is being conducted on by-products in the form of orange and grapefruit juice concentrates. Experiments in this line are being made at the plant of Bruce's Products, Inc., in Tampa, involving several thousand boxes of fruit. It is hoped that the Exchange Juice Company may develop something in this line, whereby a considerable volume of lower grades of fruit may be utilized.

The canned grapefruit contracts, mentioned in the preceding chapter as negotiated by the Exchange for the sale of canned grapefruit at 90 cents per box, was doubtless another factor of importance in the utilization of the large 1930-31 crop. Moreover, cooperative activity in the canning industry had also progressed rapidly over the preceding three years. While the independent canners fared poorly during the 1930-31 season, producer-controlled canneries made great strides. Feeling that this development of cooperative canning offered some opportunity for the Exchange, the organization established a canning division to supervise a standardized product and to effect favorable sales distribution. This action was taken in March, 1930, and by the end of May, three Florida factories were producing canned grapefruit for Exchange sale. They were the Golden Triangle Canning Corporation of Eustis, Indian River Exchange Canners of Fort Pierce, and DeSoto Canners Association of Arcadia.

The 1930-31 season thus concluded on a note of general optimism throughout the industry and of extremely bolstered confidence of the Exchange in its ability to move large volumes of fruit into profitable utilization channels. After seasons of adversity, the long predicted crop of 74,000,000 boxes had arrived and had been moved in spite of national economic conditions that were less than favorable. The big Florida industry as it is known today was beginning to emerge. Foundations for industrial fortunes were being laid, and Florida was on its way to becoming the largest citrus producer in the world.

CHAPTER 18

1931-1932

WHILE THE 1930-31 season was progressing to its conclusion, as indicated in the preceding chapter, the question of moving the headquarters of the Exchange from Tampa to some location in the interior of the state had been gaining momentum. At a series of board meetings commencing in January, 1931, this movement gained considerable support and after several stormy sessions was finally adopted as the "sense of the board." Once the board had settled this question, the problem of location of the headquarters became of paramount importance. Here again there was extreme controversy, but the decision was eventually reached that the new location would be in Winter Haven.

The decision was short-lived. As previously stated, the Federal Farm Board, through its lending facilities, had been providing vital assistance to the Florida Citrus Exchange during the economic depression of this era. Because of the Farm Board's importance in this respect, it had been asked to consider the effect of movement of Exchange headquarters to Winter Haven, and to offer its opinion of the proposed move. On June 19, 1931, the Federal Farm Board advised in a long letter to the board that "after careful consideration of all the facts developed by the survey, the Federal Farm Board has reached the conclusion that the Exchange headquarters should remain in Tampa." The advice of the highly regarded Farm Board was accepted and, although we shall discuss a change in location within the city of Tampa later, the question of moving the headquarters from Tampa inland was to remain dormant for many years.

The board of directors met in annual meeting on June 2, 1931, to organize for the 1931-32 season. Seated as directors were J. O. Carr of Charlotte Citrus Sub-Exchange, J. C. Chase of Chase and Company, W. O. Talbott of Dade, Rupert Smith of DeSoto, John A. Snively of Florence Villa, Homer Needles of Indian River, D. C. Gillett of International, J. G. Grossenbacher of Lake Apopka, John Morley of Lake Region, R. K. Thompson of Manatee, W. R. Lee of Marion, C. A. Garrett of Orange, George Speese of Polk, John D. Clark of Ridge, C. H. Walker of Scenic, A. W. Hurley of Seminole-Orange, R. J. Kepler, Jr., of St. Johns River, H. E. Cornell of Winter Haven, Marvin H. Walker of Hillsborough, and John S. Taylor of Pinellas. Special directors were F. S. Ruth, D. A. Hunt, E. L. Wirt, and L. B. Skinner. Clinton Bolick was elected as associate director. Prior to the election of officers at this meeting, the board was handed a letter of resignation from J. C. Chase, president of the Exchange. Although he had completed the tenure to which he had been elected, the letter apparently was designed to eliminate any consideration of the board in returning him to the presidency during the 1931-32 season.

As organized for this season, officials of the Exchange were John A. Snively, president; John S. Taylor, first vice-president; J. G. Grossenbacher, second vice-president; R. J. Kepler, Jr., third vice-president; and Rupert Smith, fourth vice-president. E. L. Wirt was returned as chairman of the board. Re-elected to staff positions were C. C. Commander as general manager, F. W. Davis as general sales manager, J. Reed Curry as manager of the organization department, and Judge William Hunter as attorney.

With regard to the board, at the annual meeting of the Hillsborough Citrus Sub-Exchange held on July 31, 1931, Marvin H. Walker was replaced as representative of that sub-exchange by B. E. Stall. A subsequent consolidation of the Charlotte Citrus Sub-Exchange and the DeSoto Citrus Sub-Exchange resulted in cancellation of qualifications of Rupert Smith and J. O. Carr who had represented these two organizations on the board. J. O. Carr was named, however, to represent the newly consolidated sub-exchange, and was also thereafter elected by the board to be fourth vice-president, filling the vacancy left by Smith.

The 1931-32 season saw the total American production of citrus drop below that of the preceding season, but it was considered greater than the production of any other season in the history of the industry. Further, this heavy production was marketed in the face of the worsening economic depression which by now had been described as the worst national slump since Civil War times. Low prices on Florida citrus were generally experienced throughout the season. In spite of this fact, reports indicate that average net returns to Florida citrus growers were considerably in excess of those obtained by producers of other agricultural commodities throughout the nation. For the first time citrus producers faced a new development which heralded the coming of a dangerous competitor in future marketing. Tomato juice, practically unheard of in seasons previous, was marketed in the grand total of 1,500,000 cases during the 1931-32 citrus season. Strongly supported by an intense advertising campaign conducted by packers of tomato juice, widespread consumer acceptance was becoming of major concern to the Florida Citrus Exchange as well as to all of the industry.

Another alarming factor, as evidenced by the annual report for the 1931-32 season, was the entry of chain stores into the growth, on-tree purchase, and shipment of fruit within the state. Florida citrus growers and shippers alike viewed this development with considerable alarm as adversely affecting the future marketing position of citrus fruits.

In view of the steady rise in truck shipments in current times, it is interesting to review a report by the general manager of truck operations during the 1931-32 season.

The development of trucks as carriers and their operators as nomadic and largely irresponsible sales agents for Florida citrus has been another factor of considerable importance in the movement of this season's crop. The apparent price advantage obtained by the sale of fruit to these carriers at the packing house is sacrificed in the marketing of the balance of the crop handled through standard channels.

In those markets reached by trucks filled with bulk fruit, driven day and night with low overhead, a heavy price toll is paid by the standard packs which must meet this price competition.

Government check on the truck movement of citrus from Florida was maintained this season through December. The total moved by these carriers to January 1st was 833,572 boxes, over half of which moved in the single month of December. Truck sales did not begin to decline appreciably until March of this year, and thus the total movement for the season undoubtedly is many times the total of the recorded period. The truck volume, therefore, has become a large factor in the movement of the total crop.

An examination of its widespread distribution illustrates the menace which exists unless some control is exercised. One truck load sold at low prices in an average sized market will affect negatively the prices on three cars of packed and graded fruit.

A change in Exchange policy during the 1931-32 season, although not spelled out clearly in records of that period, is evident in its operations for that season. There was an effort to concentrate all activities solely upon the promotion and sale of Exchange growers' products. Other planned activities—and they were numerous—were abandoned. The implementation of this policy, together with a reduction in estimated volume through the season, resulted in definite economies. On July 31, 1931, salaries of all employees in Florida and in terminal markets which exceeded $100 monthly were reduced by 10 per cent. An additional 5 per cent reduction on all salaries above $125 monthly became effective on April 1, 1932. Rearrangements of departmental activities where adjustments could be made without loss of efficiency and the discontinuation of many activities served to cut operational costs to bare minimums. It should be remembered that the depression was at its worst during this period and that every conceivable economy was placed in effect to bring back to the grower the absolute maximum price for his fruit.

Export volume for the season, which dipped to approximately 19 per cent under the 1930-31 season, was caused mainly by Great Britain's decision to abandon the gold standard in October, 1931. The English pound dropped from par to figures as low as $3.28 as a result of this move, and further difficulties were encountered when the British Parliament enacted a 10 per cent tariff on all citrus imports. Nonetheless, export shipments to the United Kingdom totaled 500,000 boxes during the season.

The consumer-bag container business also dipped during the 1931-32 season so far as the Exchange was concerned. Developments in the bag industry, according to available reports, eliminated the possibility of a continuation of the exclusive development of the bag business by the Florida Citrus Exchange. With other operators coming into the bagged-fruit picture, Exchange movement dropped from 705 cars in 1930-31 to less than 150 cars in 1931-32, a decrease apparently caused by the low competitive prices resulting from relaxation of the high standards established by the Exchange. Owing to what appears to have been a general tendency to downgrade the bagged package with inferior fruit, the Exchange could not compete in the matter of price and still provide price protection to its growers.

In developing a profitable market for off-sized fruit of member growers, some Exchange associations had built or leased and operated canning plants by this time. Under such arrangements, the Exchange acted as sales agents for these production groups much in the same capacity that it functioned in fresh fruit.

During the 1931-32 season, the Florida Citrus Exchange through its canning division handled sales for four units. These were located at Lake Wales, Arcadia, Bradenton, and Tampa. Combined production of these four units reached 133,377 cases of canned grapefruit hearts and juice. Wide distribution of SEALD SWEET canned grapefruit juice was slowly being organized through an effective brokerage system.

CHAPTER 19

1932-1933

THE 1932-33 season was, from all available indications, a season of almost constant change in policy and in conduct of the business of the Florida Citrus Exchange. As the previous season had drawn to a close, recommendations had been received by the board from a committee of its own members, as well as from the Erwin, Wasey advertising agency, that several fundamental changes in the organization of the Exchange be effected. These recommendations spelled out a feeling of the need for a reduction in the size of the board, for a further withdrawal of the Exchange from all activities other than the sale of fruit for its members, and for a method of eliminating duplication of activity in the general offices and in the sub-exchange offices. Some of these changes will be discussed in the following pages of this chapter; however, in the interest of continuity, important developments attending the seating of the board of directors for the 1932-33 season should first be reviewed.

It had become apparent prior to the annual meeting on June 9, 1932, that John A. Snively, president of the Exchange during the 1931-32 season, would not make himself available for re-election to that office. His reasons are outlined in a letter of resignation, the complete text of which is recorded in the minutes of the annual meeting. We quote from a part of that letter:

I want to repeat that I am not a candidate for this office and believe that, for the best interests of this organization, I should not be elected. I do not possess a political nature and am, perhaps, too forceful in expressing my views. I have at no time meant to be rude, but I have naturally hurt some of you. Therefore, I want to assure the Board at this time that I have no personal feeling against any member in writing this letter, as I believe that the Board honestly endeavors to do the best it can for the grower members of the organization.

I have an additional reason for my refusing this honor. With a definitely established policy, the Exchange could not have a salaried President. I have always felt that the President of the Exchange should have enough stake in the industry to be willing to serve without a salary and that the industry should hire any talent it needs for specific purposes.

Still another reason for my refusal is the fact that I am from Polk County. The President of the Exchange has come from Polk County most of the time. I know that that fact, together with other ideas which have originated in Polk County, has engendered a great deal of feeling in the Exchange. Certain prejudices have been fostered which should not be allowed to exist.

I further believe that the President of the Exchange should be elected by as nearly unanimous a vote as possible. Certainly he should be a man whom nobody can charge with having selfish motives in holding the office.

There are four well-organized territories in the Florida Citrus Exchange—Polk County, Orange County, the Indian River Sub-Exchange, and Pinellas County. I believe that the next President of the Exchange should come either from Orange County, Pinellas County, or the Indian River Sub-Exchange. Beyond this, however, he should come from a well-organized Association and should have a substantial interest in the citrus business.

My personal desire is not to be on the Tampa board at all. I could thus demonstrate to the growers of this state that a grower who has been active in Exchange affairs can become a grower and still have confidence that he will be treated properly by the Florida Citrus Exchange, as well as having an effective voice in the management of its affairs through the local Association and Sub-Exchange.

Directors seated at the June 9, 1932, meeting were William Edwards of Zellwood, J. C Chase of Sanford representing Chase Sub-Exchange, Frank G. Clark of Indian River City representing Dade and Indian River Sub-Exchange, J. O. Carr of Fort Ogden representing DeSoto, John A. Snively of Winter Haven representing Florence Villa, C. B. Hipson of Umatilla representing Lake and

Marion, B. E. Stall of Tampa representing Hillsborough, L. L. Lowry of Orlando representing International, J. G. Grossenbacher of Plymouth representing Lake Apopka, John Morley of Lake Alfred representing Lake Region, Lee S. Day of Bradenton representing Manatee, C. A. Garrett of Kissimmee representing Orange, John S. Taylor of Largo representing Pinellas, Vet L. Brown of Bartow representing Polk, J. K. Stuart of Bartow representing Ridge, C. H. Walker of Bartow representing Scenic, A. W. Hurley of Winter Garden representing Seminole-Orange, R. J. Kepler, Jr., of DeLand representing St. Johns River, H. E. Cornell of Winter Haven representing Winter Haven, and Clinton Bolick of Fort Myers representing Lee.

Officers and key department heads of the Exchange elected on June 9, 1932, for the 1932-33 season were William Edwards, president and chairman of the board; John S. Taylor, first vice-president; Frank G. Clark, second vice-president; R. J. Kepler, Jr., third vice-president; and J. O. Carr, fourth vice-president.

C. C. Commander was returned as general manager; F. W. Davis, sales manager; Harold Crews, manager of the field department; E. E. Lines, advertising manager, E. D. Dow, traffic manager; A. B. Steuart, comptroller; W. T. Covode, cashier; O. M. Felix, secretary; and Judge William Hunter, attorney.

With regard to the 1932-33 season itself, a report from the general manager covering its complications is enlightening:

The season 1932-33 has presented a most pathetic picture from a marketing standpoint. For the first time in history, food sales per capita dropped heavily. A decrease of approximately 17 per cent was experienced during the year.

The general features of the depression—unemployment, business stagnation, paralysis of credit facilities—are altogether too well known to require discussion. Suffice it to say that these conditions affected the marketing of Florida citrus during the 1932-33 season more than ever before in the history of the state.

Sales were not decreased as occurred in the case of staple food products, but these causes made themselves evident with increased effectiveness in the returns realized. Disastrous competition with other food products loaded onto markets of starving buying power at sacrifice prices added a final touch to the already difficult situation.

Added to the "difficult situation" outlined by General Manager Commander were other problems probably more inherent within the industry itself. Again in the 1932-33 season, the average quality of the state's citrus crop was apparently subnormal. The increased activity in what was, of course, a "buyers' market" by the speculative interests, particularly the chain stores, worked to bring cooperative interests to a virtual checkmate. Truck shipments of unregulated, unstandardized quality continued throughout the season.

Another difficulty is brought forth by General Manager Commander in his annual report for the 1932-33 season:

> The chief difficulty with which the industry is handicapped, however, remains its utter inability to regulate its volume of shipments from day to day or week to week in line with what the market can take, at a price representing a fair return to the producer. The law of supply and demand cannot be violated without penalty. If the supply of Florida citrus exceeds any given demand, an adverse price reaction immediately is experienced.
>
> This economic law was completely disregarded in the movement of the 1932-33 citrus crop. Heavier supplies were shipped to markets from the state, regardless of the fact that the price was consistently downward.
>
> Nature produces on Florida citrus groves a greatly variable supply. The demand for this supply fluctuates with economic conditions and the effort put behind its stimulation.
>
> Some outside force, therefore, must be exercised to coordinate the factors of supply and demand before the economies of the situation do so at a tremendous cost to the producers.

Apparently in recognition of this situation, responsible leaders within the Florida citrus industry at that time attempted what the Exchange felt to be a most sincere and wholehearted effort to apply profitable control in regulating grapefruit shipments, the prices of which were already well in red ink. All control was turned over to a representative committee of five individual citrus men who were given full authority to act. They represented an estimated 85 per cent of the grapefruit tonnage in the state and had the complete support of that percentage of the producers. Texas and Puerto Rican crops were off the market, leaving to Florida a virtual monopoly on grapefruit.

Although all circumstances seemed apparently favorable to this attempt at regulation for the benefit of producers, it is obvious that the plan failed. The 15 per cent of producers not subscribing to the plan flooded the markets with unregulated shipments of grapefruit as prices began to rise. At any rate, the markets were broken after an initial substantial rise had been obtained.

But the industry did not give up the idea of regulated distribution and turned next to the state legislature for relief. It presented two bills for enactment by that body. The first of these proposed the creation of an agricultural commission, which would act upon petition of 75 per cent of any commodity, and would enforce 100 per cent cooperation in its regulations. A second bill proposed definite regulations and control of standards. These bills, with considerable merit in both, were not enacted. General legislative instability caused by the tremendous state-wide economic difficulties seems to have doomed the citrus measures rather than organized opposition.

Reporting to his board on the death of the two proposals, Commander said, "Florida citrus growers, however, need not abandon hope for the effective application of these principles. The Federal government has considered the grant of similar controls in other industries operating nationally and in similar difficulties. With the failure of this state legislation, the industry should seek the cooperation of the Texas and California citrus industries in contacting the Federal government for the enactment of these principles into national legislation."

Although the final entry of Florida's citrus industry into the Federal Marketing Agreement did not become reality until May, 1936, the necessity for control of shipments was at this time clearly apparent. There is little doubt that the ideas expressed above represent fairly the viewpoint of the Exchange and subsequent support of the marketing agreement.

With the failure of the industry to regulate successfully its shipments with as much as 85 per cent control, the Exchange was convinced that it had perhaps placed too much emphasis in the past on the procurement of tonnage control. This feeling was the basis of one of the most radical and effective policy changes in the history of the organization. All effort to obtain additional tonnage by grower persuasion, or by purchase, was abandoned. Activities not relating directly to the sale of grower-members' fruit were eliminated. The entire organization was placed on a basis competitive with other shippers in the industry.

The natural advantages of superior volume and at-cost operations were quickly adopted by the Exchange as a plan to make possible a far more favorable future development in tonnage. The theory of "dominant control" was forgotten. Competitive retain figures, lower than those of any other responsible sales agency in the state, were instigated, and the Exchange's merchandising and advertising programs were tailored to fit into the reduced retain figures. Sub-exchange operations, which had developed into an extremely complicated and cumbersome system through the years, were absorbed into the parent organization, which would hereafter deal directly with the associations in securing fruit on order. Great time and expense factors were saved in this reduction of force, and it was the concensus of the existing board members that the realignment would result in greater efficiency throughout the sales system.

It seems certain that the difficulties experienced in the marketing of the 1932-33 season were crystallized into an appreciation by the board of the necessity for changes in the basic structure of the Florida Citrus Exchange. These difficulties forced what was generally considered to be improvements in the entire Exchange system. It is interesting to note, too, that the depression produced another reaction which was to be of benefit to the industry for several years. The insistent demand for a substantial reduction of the money spread between the delivered price and the grower's return on the trees had fostered the development of a large-scale increase in the use of water transportation in the movement of citrus. During the 1931-32 season, water transport facilities carried only about 7 per cent of the total citrus crop. Economic necessity increased this total to nearly 23 per cent during the 1932-33 season, with most of the loss falling to the railroads.

Along with the realignment of the Exchange system, the field and organization departments were eliminated late in the 1932-33 season. The Exchange's inspection service was continued under the direct supervision of the general manager. And with the elimination of sub-exchange offices, the state was divided into four districts headed by district managers in the employ of the Tampa office. Generally, these districts included the northern half of the citrus belt in District 1, the central section of the belt in District 2, the west coast citrus areas in District 3, and the east coast citrus producing area in District 4. While there were many other changes in the operational structure of the Exchange during this period, they were mostly departmental adjustments made necessary by those changes already outlined in some detail.

The most revealing aspect of the difficult 1932-33 season was the unrelenting pressure applied by economics for changes in the entire concept of the Florida Citrus Exchange. It must be remembered that prior to the 1932-33 season, the Exchange had grown into a cumbersome organization with fingers in almost every phase of the industry. What it had become was the result of the sometimes careful, sometimes tempestuous, always enlarging action of its boards based on actions of prior boards. It is unlikely that anything short of the economic pressures of the times could have deployed this inertia so effectively as to have caused almost complete reversals of idea and opinion that had been built up within the Exchange during the course of the preceding twenty-three years.

CHAPTER 20

1933-1934

THE ANNUAL meeting of the Florida Citrus Exchange for the 1932-33 season took place on June 8, 1933. The board of directors for the 1933-34 season as finally seated included J. C. Chase of Sanford representing the Chase Sub-Exchange, Frank G. Clark of Indian River City representing Dade and Indian River sub-exchanges, J. O. Carr of Fort Ogden representing DeSoto Sub-Exchange, John A. Snively of Winter Haven representing Florence Villa Sub-Exchange, C. B. Hipson of Umatilla representing Lake and Marion sub-exchanges, B. E. Stall of Tampa representing Hillsborough Sub-Exchange, L. L. Lowry of Tampa representing the International Sub-Exchange, J. G. Grossenbacher of Plymouth representing Lake Apopka Sub-Exchange, Charles B. Anderson of Tampa representing Lake Region Sub-Exchange, Lee S. Day of Bradenton representing Manatee Sub-Exchange, C. A. Garrett of Kissimmee representing Orange Sub-Exchange, John S. Taylor of Largo representing Pinellas Sub-Exchange, George W. Mershon of Lakeland representing Polk Sub-Exchange, J. D. Clark of Waverly representing Ridge Sub-Exchange, L. T. Farmer of DeSoto City representing Scenic Sub-Exchange, A. W. Hurley of Winter Garden representing Seminole-Orange Sub-Exchange, Francis P. Whitehair of DeLand representing St. Johns River Sub-Exchange, and H. E. Cornell of Winter Haven representing Winter Haven Sub-Exchange.

Officers and department heads elected for the 1933-34 season were John S. Taylor, president and chairman of the board; John A.

Snively, first vice-president; Frank G. Clark, second vice-president, A. W. Hurley, third vice-president; J. O. Carr, fourth vice-president; C. C. Commander, general manager; E. E. Patterson, sales manager; Harold Crews, assistant to the general manager; E. E. Lines, advertising manager; L. D. Aulls, traffic manager; A. B. Steuart, comptroller; W. T. Covode, cashier; O. M. Felix, secretary; and Judge William Hunter, attorney.

On August 4, at its regular meeting, the board was notified of the resignation of John D. Clark as a member. D. A. Hunt was subsequently seated to replace Clark.

One interesting development in the structure of the Exchange took place on June 23, 1933, when the combined boards of the Exchange and the Growers Loan and Guaranty Company resolved to consolidate their legal departments into one expanded department for more efficiency and elimination of delay in legal matters. On September 27, 1933, the board authorized the retention of Judge William Hunter as the chief counselor until the next annual meeting of the organization. The attorney who had been acting on behalf of the Growers Loan and Guaranty Company, L. M. Turner, was to become an assistant to Hunter, and a young attorney from Sanford, Counts Johnson, was employed to serve also as an assistant attorney. Johnson, who began his tenure of service with the Exchange soon after, was to continue his association with the organization for many years. At the time of this writing, June, 1959, he remains as secretary of both the Growers Loan and Guaranty Company and the Florida Citrus Exchange and acts as legal counsel for all Exchange activities.

During the 1933-34 season the Exchange apparently began to shake off the results of the declining depression. Celebrating its Silver Anniversary during this season, the Exchange could take note of more favorable circumstances both within the industry and within the organization. A Silver Anniversary brochure published by the Exchange had this to say:

In its 25th year the Exchange passed a quarter-century total of the first hundred million boxes of Florida oranges, grapefruit, and tangerines sold through the far-reaching facilities built up through the years to provide wider markets for the fruit of Exchange grower-members.

Back of any enduring organization will be found a substantial reason for its existence. The Exchange was conceived to render a needed service, under appropriate grower control and ownership.

Following its first year of pioneering in the industry, the Florida Citrus Exchange found ample opportunity for the splendid energy of its leaders. Many of the accomplishments of the Exchange, begun in the interest of its grower-members, have resulted in widespread benefit to the Florida citrus industry as a whole.

The brochure, in a summary of highlights of the twenty-five years, cited the sale of the 100,000,000 boxes at a gross return of $361,000,-000. It acclaimed the development of sales promotion for SEALD SWEET and MOR-JUICE brands through sixteen Exchange Offices in Northern markets, and proudly noted that it had the highest type of representation in 145 other markets in 46 states. It pointed out that the Exchange was also ably represented in overseas export markets. Total advertising effort on brands alone was listed as nearly $500,000 during the years preceding, and the brochure claimed a total roster of growers in excess of 6,000 members. It concluded with this summary:

Thus, the ideals of leaders who visioned the Exchange 25 years ago, have come true in fact today. Thus Exchange service marches on toward a greater tomorrow. And what better test of the worth of the Exchange service could there be than the fact that in all five districts of the state, greatly increased Exchange tonnage is reported for 1934-35. The second hundred million is on the way.

The district manager plan, mentioned in the preceding chapter, had developed by September, 1933, to the extent that three managers had already been appointed by the board. They were L. A. Hakes in District 1, O. J. Harvey in District 3, and G. R. Brooks in District 4.

On November 17, 1933, the board had asked the Farm Credit Administration for an extension of credit terms involved in loans that had been utilized for mergers and outright purchase of packing facilities in some cases during the depression. In a resolution on that date, the board asked that the Farm Credit Association study the Exchange's audits and balance sheets as a prerequisite to granting additional time for the repayment of certain phases of these loans. It also asked that the study include a survey of all Exchange policies and operations, and make suggestions as to how the operational procedures of the cooperative could be improved.

The result of this study was handed to the board at its meeting of April 12, 1934, in the form of a lengthy letter written by J. W. Jones, a representative of the Farm Credit Administration. While much of the letter dealt with specific details not important to the history or progress of the Exchange, one section of the paper seems to represent an unbiased description of the organization's marketing problems of this period. This section of the letter noted:

The Florida Citrus Exchange, like any other marketing agency, is faced with the problem of adjusting its methods to constantly changing conditions. Probably no other five-year period has seen marketing conditions more rapidly shifting and drastically changed than those witnessed during the last five years. A declining price level which made buyers of fruit unwilling to assume the risk of the price change between loading the car and arrival in the market has affected f.o.b. sales and increased the percentage of citrus fruit going to auctions. The credit stringency in Eastern markets has also contributed to this change.

The increasing use of motor trucks and boats for the transportation of fruit has almost revolutionized the method of doing business in some markets. But the motor trucks and boats are apparently here to stay and means and methods must be found to take advantage of these means of transportation. The produce trade in metropolitan markets is making some adjustments and will make still further adjustments to meet these conditions. One of the most serious marketing problems affecting the Florida Citrus Exchange has arisen in the increase of the use of motor trucks and shipment of fruit without packing in the standard container.

It seems to me that the evils of the truck have arisen because the transportation and merchandising are being performed by the same agency—the trucker. Similar confusion would exist in the marketing of boxed fruit if the railroads serving metropolitan areas purchased fruit in Florida, transported it to market, and they themselves engaged in the merchandising of it. It seems that the Florida Citrus Exchange would lessen the confusion caused by motor trucks if it would itself do the merchandising in Southeastern markets and employ trucks as a transportation agency for the service of transportation only.

Some order might be put in the bulk-fruit movement in the Southeastern markets if this policy was inaugurated. It seems that opinion is divided as to the merits of shipping bulk fruit. On the other hand it seems the movement of fruit in bulk into some markets has become an established practice. Some degree of control of the movement of bulk fruit may be established before another season through the Agricultural Adjustment Administration and the Florida Control Committee.

The Exchange should develop methods of selling that will enable it to sell to any agency that will pay the highest price for fruit, and let the purchaser select the method of transportation and the container so long as the price of purchase is in line with the price that may be secured through other methods of distribution. Of course, the coordination of truck and bulk sales will be necessary to prevent price chiseling and unequal bargaining. It seems inadvisable to attempt the fixing of a uniform price on all sales to trucks because of the differences in the quality of fruit. For instance, a uniform price is not even secured on the New York auction. On the other hand local packing houses should work closely with district managers and the sales department on all sales that involve transportation by truck.

At the regular meeting of the board on May 18, 1934, an exchange of communications between the Exchange National Bank of Winter Haven and the Florida Citrus Exchange discussed the not inconsiderable assistance given by the Exchange to the federal government for use in the establishment of the Federal Citrus By-Products Laboratory at Winter Haven.

April 24, 1934

Florida Citrus Exchange
Tampa, Florida

Gentlemen:

In August, 1931, you kindly contributed to a $6000.00 fund to be used in furnishing quarters, including building and grounds, for a Federal citrus by-products laboratory, and these funds were paid in to this bank as trustee and we hold the title to this property in trust for the various contributors to the fund.

The Federal Government established the laboratory and has been making splendid progress in research work in citrus by-products. We are anxious to have the Government extend this work and we are advised that this may not be done unless the Government itself owns the property, and this letter is for the purpose of requesting permission for this bank, as trustee, to deed to the U.S. the property now being used as a citrus laboratory.

Will you please consider this matter, and if agreeable to you, will you so advise us in writing giving permission that this bank, as trustee of the fund, may deed the same to the U.S. by Special Warranty Deed?

We will greatly appreciate your immediate attention, and awaiting your response, we are

Respectfully yours,
/S/ L. B. Anderson
L. B. Anderson
Vice President,
Exchange National Bank of
Winter Haven

A resolution of the board of directors, Florida Citrus Exchange, was returned to the Exchange National Bank of Winter Haven granting the permission requested in the foregoing letter:

WHEREAS, the Florida Citrus Exchange, in October 1931, appropriated and contributed the sum of Two Thousand Dollars to a Six Thousand Dollar fund to be used in furnishing quarters, including building and grounds, for a Federal citrus by-products laboratory, and such funds were paid to Exchange National Bank of Winter Haven, Winter Haven, Florida, as trustee, which trustee holds the title to the property purchased with such funds in trust for various contributors, and

WHEREAS, the Federal Government has heretofore established such laboratory and has made splendid progress in research work in citrus by-products, and it is now expedient and advisable that the Federal Government extend this work, and

WHEREAS, it is understood that no appreciable extension of such research work can be done unless the Government itself owns such property, now therefore

BE IT RESOLVED BY THE BOARD OF DIRECTORS OF THE FLORIDA CITRUS EXCHANGE IN REGULAR MEETING DULY ASSEMBLED:

That the Florida Citrus Exchange does hereby consent to Exchange National Bank of Winter Haven, Winter Haven, Florida, as trustees, conveying to the United States of America, or its nominee, by proper special warranty deed, all of the aforementioned property and quarters, including buildings and grounds, situate in Winter Haven, Florida, and said Florida Citrus Exchange does hereby renounce and disclaim in favor of the United States of America, any and all right title and/or interest in and to said property and every part thereof.

That the President and Secretary of the Florida Citrus Exchange be and they are hereby authorized and empowered to make and execute such written instrument or instruments as may be necessary to enable and fully authorize said Exchange National Bank of Winter Haven, as trustee, to transfer title to such property to the United States of America, or its nominee.

The closing days of the 1933-34 season brought with them the resignation from the board of John A. Snively of Winter Haven. A member of the board of Florence Villa Citrus Growers Association for twenty-one years, and a director of the Florida Citrus Exchange for twelve years, Snively had also served one full term as president of the Exchange. While his reasons for resignation are not officially recorded, a resolution of high esteem for him was documented by the board in the official minutes of the organization. W. C. Van Clief of Winter Haven was later named by the Florence Villa Citrus Growers Association to the vacancy left by Snively.

It is interesting to note, in a review of the activities of the Exchange during the 1933-34 season, that a resolution appears in the records of the annual meeting on June 7, 1934, with regard to the discontinuance of the Exchange in all grapefruit canning interests:

> Be It Resolved. . . that the marketing of canned grapefruit and canned grapefruit juice by the Florida Citrus Exchange be and the same is hereby discontinued, and that hereafter none of the operations of any cannery engaged in canning grapefruit and/or canning grapefruit juice shall in any manner whatsoever, directly or indirectly, be financed by the Florida Citrus Exchange.

Thus came the full cycle of the interests of the Exchange which began with fresh fruit, slowly extended its interests into virtually all phases of research, and into the canning of citrus, and finally returned to the exclusive interest of the marketing of fresh citrus fruit for its members.

CHAPTER 21

1934-1935

THE 1934-35 season was to be for the Exchange, as well as the entire Florida citrus industry, a period of economic conflictions and alternating hope and despair. But it did see the enactment of state legislation establishing the Florida Citrus Commission which, in the final analysis, was to set the stage for a period of unheralded citrus prosperity which has continued to present times.

Because it had much to do with the progress of these developments, the board of directors of the Florida Citrus Exchange during the 1934-35 season was to share responsibility with other important industry representatives in laying a course for the future. The board included Frank G. Clark of Indian River City, J. O. Carr of Fort Ogden, W. C. Van Clief of Winter Haven, C. B. Hipson of Umatilla, B. E. Stall of Tampa, L. L. Lowry of Tampa, J. C. Palmer of Windermere, Charles B. Anderson of Tampa, Lee S. Day of Bradenton, C. A. Garrett of Kissimmee, John S. Taylor of Largo, George W. Mershon of Lakeland, D. A. Hunt of Lake Wales, C. H. Walker of Bartow, A. W. Hurley of Winter Garden, Francis P. Whitehair of DeLand, and H. E. Cornell of Winter Haven.

President of the Exchange during the season was John S. Taylor. Frank G. Clark was first vice-president; A. W. Hurley, second vice-president; J. O. Carr, third vice-president; L. L. Lowry, fourth vice-president; C. C. Commander, general manager; E. E. Patterson, sales manager; Harold Crews, assistant to the general manager;

Earl Lines, advertising manager; L. D. Aulls, traffic manager, S. L. Looney, treasurer-comptroller; W. T. Covode, cashier; O. M. Felix, secretary; and William Hunter, attorney.

With regard to the season itself, the Florida citrus industry at first faced a most difficult marketing situation as the harvest commenced in September. The estimated volume of the Florida crop was approximately 36,000,000 boxes. Grapefruit volume in Texas was expected to double the 1933-34 volume with an estimated 3,000,000 boxes. California was expected to produce an unprecedented volume of oranges, both Navels and Valencias. Altogether, experts could foresee nearly 100,000,000 boxes of citrus being marketed, a volume that had never before been reached in the citrus history. This situation, of course, presented a formidable marketing problem, and records indicate that the outlook for the Exchange was a gloomy one.

The early part of the season was difficult for all Florida citrus as had been expected. The Federal Marketing Agreement was eventually declared effective on December 18, which under normal circumstances would have been cause for optimism within the industry. The Marketing Agreement, however, was preceded one week by the disastrous freezes of December 11 and 12, 1934. The freezes, of course, made altogether impractical any assistance through the Agreement for the industry.

Florida citrus was not alone in its troubles. Texas suffered from windstorms and from a series of critical freezes. Palestine had severe losses and Spain experienced one of its most disastrous freezes of all time. But citrus difficulties in Florida during five years of economic depression, hurricanes, freezes, and Med-fly losses made the situation more serious in Florida, perhaps, than in any other citrus-producing area. But in the face of all these handicaps, Florida managed to ship a volume of citrus during the season nearly equal to that shipped during the previous season. The Exchange made gains in membership and acreage which increased its total annual volume by an estimated 4 per cent over the 1933-34 season, despite the withdrawal of the large Chase and Company organization from the Exchange.

The increasing production of citrus fruits in Florida, as well as in other producing areas, combined with the still retarded purchasing power of the nation, seemed to indicate that the Florida citrus industry must adopt and adhere to a more rational and businesslike handling of marketing problems. The problem inherent in the increased production of citrus had become urgent. The moment for decisive action was at hand—a situation recognized by a great portion of the industry. Coordinated effort in shipments and in merchandising, expansion of markets, and the development of consumer demand for Florida citrus could no longer be prolonged.

In an atmosphere of adversity, the Florida citrus industry exhibited unprecedented cooperative logic in turning to the state legislature for the answer to its pressing problems. For its part in the development of the citrus legislative program for the 1935 session of the state legislature, the Florida Citrus Exchange appointed a committee of its board to assist its president, John S. Taylor, and general manager in cooperation with other industrial factors in the formulation of a legislative program for the improvement of the citrus industry. Appointed to this committee were C. H. Walker, H. E. Cornell, and Frank G. Clark. The Exchange's committee was appointed on February 15, 1935, and immediately joined with a committee of Associated Growers and Shippers of Florida to draft legislative proposals. The Associated committee included L. P. Kirkland, Barney Kilgore, N. H. Vissering, W. H. Mouser, and H. C. Case.

Placing itself at the service of the joint committee was a platoon of attorneys including Judge Spessard L. Holland of Bartow, Judge O. K. Reeves of Tampa, Judge William Hunter of the Exchange, Governor Doyle Carlton of Tampa, Counts Johnson of the Exchange, C. E. Ware of Clearwater, and E. G. Grimes of Palmetto.

The various meetings held by the joint committees, according to records still available, were attended by prominent citrus growers and shippers throughout the state including C. C. Commander, R. B. Woolfolk, W. J. Howey, and Latt Maxcy. Meetings were held at Lakeland on February 26, 1935, at Tampa on March 4, 1935, at Lakeland on March 6, 1935, at Lakeland on March 20, 1935, at Tampa on March 27, 1935, at Lakeland on April 1, 1935, and again at Lakeland on April 18, 1935. At these meetings were prepared bills known as Florida Citrus Commission Act, Licensing and Bonding Act, Growers' Cost Guarantee Act, Citrus Maturity Act, Advertising Acts, Field Box Act, Color-Added Act, and the Prorate Act. Once drafted and approved by the joint committee, the legislative proposals were taken to Tallahassee where a joint house and senate citrus committee public hearing was scheduled for April 11,

1935. The hearing was attended by a large number of prominent citrus growers and shippers from all sections of the state, including all members of the joint committee and many other leading business men who were vitally interested in the welfare of the Florida citrus industry.

On June 6, 1935, the Exchange's legislative committee reported to the board with the following communication:

Supplementing a previous report made to this Board on April 19, your Citrus Legislative Committee wishes to file the following information in completion of its work which has extended over a period of four months, necessitating considerable time and travel, and commanded the full cooperation of many factors throughout the industry and the state.

Eleven preliminary meetings were held in the drafting of the legislation. Some of these lasted all day. They were open to all growers and many prominent growers attended various sessions and participated in the discussions, which finally resulted in a unanimous presentation.

Six weeks' work was necessary at Tallahassee.

Nine bills in all were drawn for presentation to the state legislature as follows:

1. A stabilization measure with provision for a State Citrus Commission.
2. Maturities Act.
3. Advertising (three bills).
4. Licensing and Bonding Citrus Fruit Dealers.
5. Field Box Act.
6. Control over Color-Added Applications.
7. Growers' Cost Guaranty Bill.

The Florida Citrus Commission will have a wide scope of authority for the stabilization and protection of the reputation of the Florida citrus fruits, on inspection and grading, all under the Commissioner of Agriculture of the State of Florida. The Commission will consist of eleven practical citrus fruit men of Florida, appointed by the Governor to establish grades and identifications and to cause prosecution for violations of these provisions. The Commission will be required to make an annual report to the Governor, and all rules, regulations, and orders by the Commission must be proposed in advance of the date upon which they become effective and are subject to protests and hearings.

The citrus maturity act makes it unlawful to sell or transport, except in the case of tree sales, any citrus fruits between the 31st of August and the 1st of December of any year unless such fruit is certified to be mature and a certificate of inspection and maturity provided, according to the rules and regulations of the State Commission. For grapefruit, the Act specifies the required juice content for all sizes. For oranges, the ratio of total soluble solids to anhydrous citric acid must not be less than 8 to 1; for tangerines 7.5 to 1. Vendors or shippers will pay the inspection fee of one cent per box, if that be required for the service, otherwise a lesser amount to be determined by the Governor and Commissioner of Agriculture. Provision is made for extending maturity inspection beyond the date, December 1st, during the seasons of late maturity or irregular bloom.

Three separate bills are proposed on advertising: one for oranges, one for grapefruit, one for tangerines. The Preamble of this Act states that it is to conserve and promote the prosperity and welfare of the industry by promoting the sales of oranges, grapefruit, and tangerines, through publicity, advertising, and sales promotion campaigns to increase the consumption of such fruits. The State Commission is empowered to set up the machinery and make contracts for carrying out the provisions of this Act which involves assessments in the form of an excise tax upon each box of tangerines at five cents.

The fourth bill provides for the licensing and bonding of citrus fruit dealers through the State Commissioner of Agriculture. License fees are established and penalties for violations are prescribed. Under the Licensing and Bonding Bill, all operators, shippers, and buyers, including trucks, will be required to keep records and to carry out contracts, operating under the rules and regulations of the State Commissioner.

The Field Box Act is for the protection of owners of registered containers and sets up a procedure whereby through proper filing of identification marks for such containers, then the same become prima facie evidence of ownership. It would be unlawful under this Act to possess the containers of another person or corporation.

The bill on Color-Added provides for state control to coordinate with provisions of the Pure Food and Drug Act under the Agricultural Department. It would thus become unlawful for any person to use or apply coloring matter to citrus fruit unless same were in accordance with state and federal requirements.

A Grower's Cost Guaranty Bill provides that under an emergency, growers may petition for the determination of the cost of production which cost would be guaranteed the grower on his sales. This bill, patterned much after the New York Milk Bill, may prove to be the means of stabilizing prices at levels that would prevent red ink shipments and red ink investments in production of citrus.

The above program makes possible the many fundamentals of merchandising, such as advertising, standardization, ethical practices, etc., which the Florida Citrus Exchange has advocated for many years. Now, the growers of the state are assured of the legal machinery for improving many marketing ills. Under this legislation, the advertising of Florida citrus becomes a reality, and also becomes an equal responsibility upon all, just as the benefits will accrue to all. Markets can be developed and consumers can be acquainted with the seasons and with the values in Florida citrus. This will reinforce our own advertising and merchandising considerably and improve the relative position of Florida citrus in the markets.

Steering these bills through both houses of the legislature was no easy matter. First, following their introduction in the Senate, came the plea for a public hearing which was held and which was attended by your committee and representative Exchange growers. There was some opposition to some features of the various bills.

In the House various amendments were proposed and it appeared for a few weeks as though the program might be defeated.

The amended bills went to Conference where after long debate and consideration an agreement was finally reached near the close of the legislative sessions, but not until much work had been done to acquaint the members that the proposed program was a balance and a much needed one in Florida.

In final amended form all bills passed, and are now in progress through the Governor's hands for signature and appointment of the Stabilization Commission. Such amendments as now appear with the bills do not, in our opinion, alter the original purposes and functioning.

We now commend the legislative program to you as an accomplished fact and ask to be discharged.

C. H. Walker, Chairman
H. E. Cornell
Frank G. Clark

In the opinion of this writer, it is doubtful if any of the primary interests in the activities prior to and during the enactment of the Florida citrus legislation of 1935 could have realized at that time the exceptional importance of what was to become the Florida Citrus Code in meeting many of the problems of the Florida citrus industry.

This era cannot be passed over without mention of the death during the season of a dedicated Exchange employee. Judge William Hunter, who had served as the Exchange's legal counsel since shortly after the organization of the cooperative, died on May 25, 1935. Judge Hunter's death was followed in July by the death of William Edwards, long-time member of the Exchange who had served as its president during the 1932-33 season.

As the complicated 1934-35 season drew to a close, it is obvious that signs of better times were apparent to those associated with the Exchange. Of importance were the slowly improving economic circumstances of the nation which heralded greater buying power on the markets. Of equal importance was the enactment of state legislation designed to correct many of the evils within the industry. It is interesting to review the text of an address by General Manager Commander at the close of the 1934-35 season:

> It really begins to look as though better days might be expected. Surely, if the fundamentals of good business are recognized and followed, Florida citrus growers will be happy again. Cooperation and coordination are necessary, however.
>
> Increased efficiency and greater economies will accrue to the Florida Citrus Exchange with the continuation of harmonious relations and efforts. With everyone pulling together, with shoulders to the wheel in a combined effort to build up and to improve our situation as growers, the Florida Citrus Exchange will progress steadily and will be thus better able to continue its fight for the growers of the industry.

To the many active citrus leaders of today who were involved in the happenings of the 1934-35 season the thought must come now that they were present at the conclusion of a well-defined period in the history of the citrus industry. For with the enactment of the legislation of 1935, the rollicking, daredevil, up-and-down acrobatics of an industry in complete conflict with itself approached conclusion. A combination of the right men in the right place at exactly the right opportunity had finally succeeded in attaining a moment of cooperation. In that moment the Florida citrus industry rang down the curtain on its formative years and graduated into an era of difficult transition.

CHAPTER 22

1935-1936

WHILE the board of the Florida Citrus Exchange for the 1934-35 season was responsible for many efforts in connection with the state legislation, it fell to the 1935-36 board to correlate this legislation with the structure of the Exchange. This board, seated on June 6, 1935, included H. G. Putnam of Oak Hill, J. O. Carr of Fort Ogden, W. C. Van Clief of Winter Haven, B. E. Stall of Tampa, L. L. Lowry of Tampa, C. B. Hipson of Umatilla, J. C. Palmer of Windermere, Charles B. Anderson of Tampa, Lee S. Day of Bradenton, Charles A. Garrett of Kissimmee, John S. Taylor of Largo, George W. Mershon of Lakeland, D. A. Hunt of Lake Wales, C. H. Walker of Bartow, A. W. Hurley of Winter Garden, Francis P. Whitehair of DeLand, and H. E. Cornell of Winter Haven.

Officers and department heads for the season included John S. Taylor, president and chairman of the board; W. C. Van Clief, first vice-president; A. W. Hurley, second vice-president; J. O. Carr, third vice-president; L. L. Lowry, fourth vice-president; C. C. Commander, general manager; Harold Crews, assistant general manager; E. E. Patterson, sales manager; Earl Lines, advertising manager; L. D. Aulls, traffic manager; S. L. Looney, treasurer-comptroller; O. M. Felix, secretary; and Counts Johnson, attorney.

Sub-exchanges of the Central Exchange included Dade and Indian River, DeSoto, Florence Villa, Hillsborough, International, Lake and Marion, Lake Apopka, Lake Region, Manatee, Orange,

Pinellas, Polk, Ridge, Scenic, St. Johns River, Seminole-Orange, and Winter Haven.

During the shipping season of 1935-36, a combination of several factors served to increase the price level for Florida citrus fruits. While these conditions affected the entire Florida citrus crop favorably, active Exchange merchandising and sales facilities apparently took greater advantage of them than the average shipper. At any rate, the Exchange moved a greater volume of fruit at a considerably increased gross revenue over the state average. The improved situation within both the Exchange and the industry was apparently caused by three specific factors: better quality, improved economic conditions and increased buying power, and the coordination of industry factors in merchandising and standardization of grades. While the freezes of 1934 had sharply reduced the citrus volume for the 1934-35 season and had adversely affected quality, these factors had disappeared by the 1935-36 season. The crop as a whole was of unusually good quality, juice content, and flavor. Trade and consumer acceptance was felt favorable and demand was relatively high. By the close of the 1935-36 season, the outlook for the 1936-37 season was considered as particularly good.

From all available correspondence and records of the 1935-36 season, it is apparent that retailers, wholesalers, jobbers, and distributors in the markets were reporting a steady and healthy demand during the entire season.

The lower volume of better-quality fruit moved into markets that were experiencing continued improving economic circumstances. Midwest markets were stronger than they had been in more than five years and markets in the East were showing even greater capacities for the purchase of Florida citrus. Southern markets bought heavily of packed fruit for the first time in four seasons. The greatly improved markets were attributed to increased buying power across the nation, but coupled to this economic factor was surely the effect of the state regulations preventing indiscriminate shipping of unpacked fruit through subnormal trade channels.

Another major factor causing improved markets during the season was the result of what the Exchange considered to be better merchandising and coordination throughout the Florida citrus industry. There is no doubt that this was due to the rigid enforcement of the new state citrus legislation. An additional stimulant to sales resulted from the state citrus advertising campaign, which tended to reduce consumer sales resistance to Florida citrus that had been excessively activated by the lower quality of the immediately preceding years.

After considerable negotiation throughout the 1935-36 season, the Florida Citrus Exchange, along with other interests, succeeded in arranging a new Federal Marketing Agreement to replace the agriculture agreement of the 1934-35 season which Secretary of Agriculture Henry Wallace had terminated at the request of a small group of influential Florida citrus growers. The new agreement went into effect on May 8, 1936. Designed to help regulate the flow of fruit from Florida into its markets by making possible proration by grade, size, and volume, the agreement was generally favorably received by most of the industry. Members of the Florida Citrus Commission, under the agreement, were named as the control committee responsible for the administration of prorate provisions through the Agricultural Adjustment Administration. Properly administered, the Exchange felt that the Federal Marketing Agreement would assist materially in stabilizing markets by regulating the volume and market acceptance of shipments to the then existing demand.

Statistics of Exchange sales during the 1935-36 season show that the Florida Citrus Exchange had not only shipped more fruit and a greater percentage of the total crop, but had returned more money to its growers than all other competitive shippers in the state. Exchange f.o.b. shipments for the 1935-36 season increased by nearly 30 per cent, mostly as a result of the direct contact between the Tampa sales office and the associations, coupled with the flexibility afforded through the district manager system. Under the revised system, the Exchange found that special orders were filled far more promptly and accurately than by the old sub-exchange method, under which orders frequently were lost to competitors because of hesitancy and delay.

In the export field, the Exchange led the industry in a move to include citrus fruits in reciprocal trade agreements with Canada. An extensive brief was prepared and filed with the Tariff Commission in Washington on March 5, 1935, with a formal hearing held on March 12. On September 15 special notice of these negotiations was made, and on November 15 the new reciprocal trade agreement between the United States and Canada was signed.

Features of the new agreement were exceptionally favorable to Florida. Prior to the signing of the new trade pact, tariff duties on oranges shipped into Canada from the United States were seventy cents per box. The new agreement eliminated all tariffs for the months of January, February, March, and April. The tariff on grapefruit was reduced from one cent to one-half cent per pound. Under these revised tariff conditions, the Canadian market was reopened to American citrus, broadening the marketing potential of Florida citrus to a considerable extent. Also, almost immediately, the movement of heavy volumes of California citrus onto the Canadian market resulted in more favorable conditions for Florida on the Eastern domestic markets.

Florida citrus canneries, which had packed over six million cases of citrus products during the 1934-35 season, decreased their utilization in the 1935-36 season to around four million cases. Nonetheless, the average pack of five million cases per season for the two-year period indicated that the Florida citrus canning industry had grown to a size which would assure its permanency. With reference to the canning industry, it must be noted that much of the volume utilized by canners had been purchased from growers at prices far below the cost of production. This factor was cause for alarm and was considered by the Exchange to be of serious detriment to the industry. By the close of the season, however, canneries were paying at an improved level for canning-grade fruit. This fact alone, in a review of the situation, seems to have placed confidence of the industry in the ability of the canners to produce and market their products at prices which would compensate the grower at least at a rate equivalent to his cost of production.

The Exchange's advertising and sales promotion activities, which had continued to function under curtailed conditions during the depression years, showed evidence of returning to something near what was felt to be a desirable level. Large two-sheet posters in New York subways and in Chicago elevated and surface railway stations featured the SEALD SWEET and MOR-JUICE products throughout the winter and spring months. It was estimated that two to three million city dwellers were exposed to these posters each day. Window, counter, and floor displays were erected by Exchange dealer service men in over sixteen thousand stores throughout the nation in a rejuvenated point-of-sale program. Independent merchants and chain-store groups apparently cooperated enthusiastically with the Exchange during this period. A series of local advertising campaigns using newspapers or radio, or in some instances both media, was carried on in cooperation with the distributing and retail trade, and more than ten thousand Exchange folders, health booklets, and other pieces of literature were distributed by the Exchange during the season.

It is timely to mention in this chapter that the cooperative plan and aims of the Florida Citrus Exchange in rendering its service to growers and associations in the marketing of fruit continued to include the indispensable operations of the Growers Loan and Guaranty Company and the Exchange Supply Company. In nineteen years of service to Exchange members, the Growers Loan and Guaranty Company had by this time lent more than $31,000,000 during both good and bad times to assist Exchange members. Liquidating its credit obligations accumulated during the depression, and carrying a substantial cash balance at all times, the Growers Loan and Guaranty Company had continued to maintain the respect and support of the important financial institutions throughout the nation.

Utilizing the services of the Exchange Supply Company, the Exchange continued to bargain for packinghouse supplies and field equipment in large quantities direct from producers at lowest possible prices. These savings accrued directly to the various associations which as stock investors owned and controlled the company. In the Exchange Supply Company the cooperative idea of services at cost was being carried into the purchasing field, thus supplementing the cooperative sales activities of the parent organization.

From the foregoing it can be assumed that the annual meeting of the Florida Citrus Exchange on June 4, 1936, convened in an atmosphere of great optimism. Directors seated at this meeting were H. G. Putnam of Oak Hill, J. O. Carr of Fort Ogden, W. C. Van Clief of Winter Haven, B. E. Stall of Tampa, L. L. Lowry of Winter Haven, C. B. Hipson of Umatilla, J. C. Palmer of Windermere, Lee S. Day of Bradenton, Charles A. Garrett of Kissimmee, John S. Taylor of Largo, George W. Mershon of Lakeland, D. A. Hunt of Lake Wales, C. H. Walker of Bartow, A. W. Hurley of Winter Garden, and H. E. Cornell of Winter Haven.

Re-elected to the presidency of the Exchange was John S. Taylor; W. C. Van Clief was elected first vice-president; A. W. Hurley, second vice-president; J. O. Carr, third vice-president; L. L. Lowry, fourth vice-president; C. C. Commander, general manager; Harold

Crews, assistant general manager; E. E. Patterson, sales manager; Earl Lines, advertising manager; L. D. Aulls, traffic manager; S. L. Looney, treasurer-comptroller; O. M. Felix, secretary; and Counts Johnson, attorney. The death of President Taylor soon after the annual meeting elevated W. C. Van Clief to the position of acting president until the October 16, 1936, meeting when Judge W. L. Tilden of Orlando was named president and chairman of the board and seated as a special director.

Fred T. Henderson of Winter Haven was subsequently named by the Lake Region Sub-Exchange to replace Charles B. Anderson; R. J. Kepler, Jr., was selected by the St. Johns River Sub-Exchange as its representative on the board and John S. Taylor, Jr., was elected by the Pinellas Sub-Exchange to assume the board vacancy caused by the death of his father.

CHAPTER 23

1936-1937

ON DECEMBER 18, 1936, the board paid memorial tribute to Sales Manager E. E. Patterson, whose death a few days previously had been an unexpected sorrow to the Florida Citrus Exchange. Finding it impossible or inadvisable to name an immediate successor to fill the vacancy in the organization left by Patterson's death, General Manager Commander assumed the duties of sales manager for the balance of the 1936-37 season.

With regard to the shipping season of 1936-37, perhaps the most outstanding feature was the fact that the three major citrus producing areas of Florida, Texas, and California, shipped a record combined total volume during the Florida season. In spite of a freeze loss of 8,000 cars by California, the combined total volume of the three states reached a weighted increase of 25.4 per cent. Texas showed the greatest increase with an increase of 15,000 cars, or 213 per cent over the 1935-36 season. Meanwhile, Florida was increasing its shipments by 17,000 cars, or 30 per cent. What seems to be particularly significant in these statistics is the phenomenal increase in grapefruit shipments from Florida and Texas. While a review of the week-by-week shipments for the season indicates that some weeks reached as high as 265 per cent increase over the corresponding weeks of the prior season, the average grapefruit shipment increase for the two states was 93 per cent increase for the season—nearly double the volume of the 1935-36 season.

It seems remarkable that while these record volumes were being shipped to American markets, price averages throughout the season for Florida producers were considered satisfactory. One of the underlying causes of this situation resided in the continuing recovery of the nation's economy. Buying power was again increased, and the nation's consumers were now purchasing items that during the depression had been considered semiluxuries.

The Florida Citrus Exchange placed much of the responsibility for improved marketing conditions on what it felt to be the successful function of the Federal Marketing Agreement and the prorates available under that agreement. Also of importance, in the opinion of the Exchange, was the purchase of surplus fruit by the federal government, canners' commitments on futures, the activities of the Florida Citrus Commission, and chain-store promotional campaigns. Reporting to the board of the Florida Citrus Exchange in May, 1937, with regard to the Marketing Agreement, General Manager C. C. Commander made this observation:

> . . . consideration should be given to the fact that a Marketing Agreement for Florida, even though properly supported by effective administration of the State's own legislative program, is not a final solution to the constant efforts of the producer to gain stabilization in the industry.
>
> Florida oranges are marketed for a price which is directly affected by the total volume of oranges moved from California, Texas, and Arizona. Correspondingly, Florida grapefruit markets are dependent upon the national supply factor which is the combined production of Florida, Texas, and Puerto Rico. In each case, it is the combined volumes which, balanced with existing demand, control the price.
>
> These factors point to but one ultimate objective. To secure full and permanent benefit from a Marketing Agreement and proration or other activities available under the Agreement, it will be necessary for the nation to establish similar operations in other producing sections. These should operate in harmony and uniformly on oranges and grapefruit as commodities.
>
> It is important that Florida growers should appreciate the magnitude to which their industry has grown. They must recognize the fact that, since it has become a national institution, it must be handled nationally if it is to continue successfully and reach any semblance of stabilization for its investments.
>
> The Florida citrus industry is no longer sufficient unto itself and able to operate without regard for other producing areas. It does

not control its own markets. Jointly with these other producing sections, however, it can. To accomplish this end, the industry, with the cooperation of the Government, must press for this type of an agreement.

It is interesting to note that the government purchase of surplus commodities was of benefit to the Florida citrus industry throughout the 1936-37 season. The agency responsible for the purchase of these surpluses apparently adopted a well-founded policy and stayed with it throughout the season. As a result cannery-grade grapefruit—to name only one instance—held at an average of 31 cents on the tree. The arrangement for federal purchase of surpluses, therefore, became from all indications a strong factor in the successful results obtained for Florida citrus producers throughout the season.

There is some evidence that the industry, during the early part of the 1936-37 season, feared that a dangerous oversupply of grapefruit existed. The alarm was short-lived, however, and at the season's end Exchange officials voiced the opinion that the canning industry could have utilized every box of fruit purchased by the government as surplus, and could have paid the growers a fair price for it. By the close of the 1936-37 canning season, canners were paying the unprecedented price of $1.25 per box for oranges. There is sufficient evidence on file to support the belief that even at this price canners found it difficult to procure enough fruit to meet their increasing demands.

The continued efficiency of the Florida Citrus Commission had by this time convinced the great majority of those in citrus of the logic of the legislation of 1935. The annual report of the Florida Citrus Exchange for the 1936-37 season includes this comment:

All growers are cognizant of the increasingly unsettled conditions which existed in the Florida citrus industry during the past few years. They became so critical about two years ago that a majority of citrus growers, the Florida Citrus Exchange, and other shippers in the state participated in the drafting of state legislation designed to improve these conditions.

This block of laws provided for the adoption of many standard merchandising fundamentals and ethics of sound business procedure. These laws became applicable to the entire industry. They fairly and impartially included all growers and operators in their provisions.

That this legislation, even in its original form, has proved profitable to the industry as a whole even the most skeptical in the industry now concede. It has been helpful in creating and maintaining demand for Florida citrus. It practically eliminated bulk shipments and provided for the inspection and certification of fruit so that buyers were able to place their commitments with confidence.

The several factors noted above resulted in an extremely favorable season. It is likewise apparent that all growers benefited from the season in direct proportion to the efficiency of their sales agencies. In this respect, the Florida Citrus Exchange had much of which to boast. The Exchange had, during the season, increased considerably its capacity to serve its members at a lower cost. In addition, the cooperative had increased distribution very substantially. By May 22, 1939, the Exchange had made sales in forty-four states and six Canadian provinces, and indications are that of a total of 308 entered markets 37 per cent were considered to be new market areas.

The intensive effort of the Exchange's sales department was registered in more favorable returns. Comparative auction returns, grade for grade and size for size, for the Exchange during this season, in comparison to the state average, indicate that the cooperative sold a total of 2,542,989 boxes of citrus up to May 1, 1937. Its average return amounted to $2.74 per box, while all competitors sold 4,838,498 boxes at an average of $2.59 per box. F.o.b. sales, under a special department in the Exchange since 1932, continued to increase throughout the 1936-37 season, concluding the season with the filling of more than 4,000 orders.

The reciprocal trade agreement with Canada, mentioned in some detail in the preceding chapter, made possible a material increase in the Exchange's business in that country. The gain over the previous season was estimated at 73.9 per cent. In considering the gain, it is interesting to note that the terms of the agreement with Canada included abandonment of all tariffs on oranges during January, February, March, and April, and that no sales were made in Canada after May 20 in the 1935-36 season. In the season of 1936-37, however, sales continued well into June with the additional tonnage adding substantially to the Exchange's overall seasonal record.

Also, with regard to the export market, the effect of the civil war in Spain was of direct interest to the Florida citrus industry generally, and of particular importance to the Florida Citrus Exchange. While the war itself was expected to have no direct bearing on Florida citrus exports to Europe, it seemed certain that its effects would be felt on the domestic market. Spanish oranges were being moved to European markets during the period of Florida's heaviest shipments, but comparative price levels made it impossible to export profitably Florida fruit to accommodate the deficiencies in Spanish volume caused by the war. There was, however, a favorable aspect for Florida citrus interests. California had developed a demand for its fruit which was suitable for export, permitting it to export substantial volumes during the spring and summer. Because of this, Florida expected that California would exploit the European field to its advantage. The loss to Europe of the Spanish volume was expected to be counterbalanced by increased exports from California. Florida could foresee that this circumstance would, in turn, relieve domestic markets in direct proportion to the California exports. This would, of course, improve the domestic marketing situation for Florida.

The work of the Exchange's dealer service program, along with its continued brand advertising, was extremely effective during the 1936-37 season. Records indicate that merchandising crews made more than 25,000 calls on wholesalers and retailers during the season. Over 4,000 complete window and interior displays, using more than 60,000 pieces of display material, were installed. Finally, subway and elevated advertising, newspapers, radio, dealer tie-ups in local retail advertising, space in food shows, cooking schools, and many other devices were used to a distinct advantage.

The relatively new legal department of the Exchange had, from all indications, been extremely active during the 1935-36 and the 1936-37 seasons. Under the direction of Counts Johnson, the department had been to a large extent responsible for drafting much of the state legislation establishing the new citrus laws. The 1936-37 season saw the legal department active in the negotiations for and preparation of the Federal Marketing Agreement and the many details necessary for its successful operation. It handled negotiations and adjustments in the Florida Citrus Exchange loans under the Farm Credit Administration, and effected the legal aspects of the redistricting problems that arose with the elimination of sub-exchange offices.

At the close of the 1936-37 season the Florida Citrus Exchange had grounds for considerable optimism. Because of returning sound financial conditions, along with increasing volume, the management looked forward to a reduction in operating costs with resulting higher net returns to Exchange growers.

These, then, were the circumstances surrounding the annual meeting of the board at the close of the 1936-37 season. Held on June 3, 1937, the meeting seated as directors H. G. Putnam of Oak Hill representing Indian River Citrus Sub-Exchange, E. F. DeBusk of Gainesville representing Lake County Citrus Sub-Exchange, J. W. Sample of Haines City representing Lake Region Citrus Sub-Exchange, S. A. Whitesell of Largo representing North Pinellas Citrus Sub-Exchange, J. C. Palmer of Windermere representing Orange County Citrus Sub-Exchange, John S. Taylor, Jr., of Largo representing Pinellas Citrus Sub-Exchange, D. A. Hunt of Lake Wales representing Polk County Citrus Sub-Exchange, C. H. Walker of Bartow representing Scenic Citrus Sub-Exchange, H. E. Cornell of Winter Haven representing Winter Haven Citrus Sub-Exchange, W. C. Van Clief of Winter Haven representing Florence Villa Citrus Sub-Exchange, L. L. Lowry of Winter Haven representing International Citrus Sub-Exchange, William G. Geier of Windermere representing Plymouth Citrus Sub-Exchange, A. W. Hurley of Winter Garden representing Lake Apopka Citrus Sub-Exchange, and W. L. Tilden as a special director-at-large.

Subsequently seated on the 1937-38 Board were R. J. Kepler, Jr., of DeLand, representing St. Johns River Citrus Sub-Exchange, and W. W. Raymond of Alva, representing West Coast Citrus Sub-Exchange.

Officers and department heads elected to serve during the 1937-38 season were W. L. Tilden, president and chairman of the board; W. C. Van Clief, first vice president; A. W. Hurley, second vice-president; William M. Moseley, third vice-president; L. L. Lowry, fourth vice-president; C. C. Commander, general manager; Harold Crews, assistant general manager; L. D. Aulls, traffic manager; S. L. Looney, treasurer-comptroller; O. M. Felix, secretary; and Counts Johnson, attorney.

CHAPTER 24

1937-1938

THE OPTIMISTIC outlook of the board of directors of the Florida Citrus Exchange, and that of a great part of the Florida citrus industry, at the close of the 1936-37 season was unfortunately not well founded. The season of 1937-38 ran early and headlong into serious marketing difficulty which plagued the industry throughout the shipping season. From practically every standpoint records of the season reveal a sharp and unfavorable contrast with the 1936-37 season for Florida citrus. Returns were uncertain and relatively meager, and frequently below the cost of production.

The causes of what the Exchange considered to be the poorest season of the immediately preceding years were varied. It seems certain that some of the difficulty lay in the unfavorable whims of Nature. Another cause was the apparently floundering national economy which had only a year before shown signs of completely recovering from the depression. There can be no doubt, however, that some of the difficulty was peculiarly man-made. Fortunately, the season progressed with most citrus leaders convinced that the problems were not permanent or without hope of profitable solution.

Constantly increasing volumes of all citrus-producing sections of the Hemisphere had long been regarded by cooperative leaders as a factor which, unless moved in an orderly manner for proper distribution along with stimulated consumer demand, would prove to be an underlying cause of disaster to the Florida citrus industry.

There was almost undeniable statistical proof of the dangers they feared. The five-year annual average from 1928 to 1932 of total production in citrus areas was 63,669,000 boxes. In 1936 the harvested crop was 85,854,000 boxes, and estimates placed the total of the 1937-38 crop from all areas at nearly 100,000,000 boxes.

There was at the time no evidence of future promise of relief from these heavy citrus volumes. The bearing and nonbearing acreage being cultivated at that time in Florida, Texas, Arizona, and California reflected 17.3 per cent as nonbearing acreage, but which would constantly come into bearing stages further to complicate the volume problem. The outlook was such that the Exchange foresaw average annual production increasing by nearly 50 per cent above the 1928-32 average by 1943, unless climatic catastrophes served to curtail production.

Coincident with the high rate of development of citrus at this time was the parallel heavy production in competitive deciduous products. For example, the 1937 apple crop of 211,060,000 bushels was the largest produced in the United States in twenty-three years, and represented the peak of a constant increase during the ten-year period from 1927 to 1937. Apples were generally considered a direct competitor of citrus, particularly when high production caused low consumer prices.

The Southern strawberry crop, also felt to be a citrus competitor at the time, experienced a 30 per cent increase in the 1937 season. Cantaloupes, peaches, pineapples from Puerto Rico and Cuba, with canned tomatoes, prunes, and pineapple juice, were all in high volume and represented active competition.

Thus, bumper crops of citrus and competitive commodities combined to set the stage for the problems of the 1937-38 Florida citrus crop. And the entire production was sent to markets which were verging on a depression as critical as the one from which the country had only recently made such remarkable recovery.

This situation can be emphasized by a look at industrial America during the 1937-38 period. Steel, by January of 1938, had dropped from an index of 131 points in 1937 to 38.6 points. Cotton mill activity was down from a high of 129.2 points in 1937 to 84 in January, 1938. Freight car loadings at 80.8 in 1937 had dropped to 62.1 in January, 1938, and levelled at 60.6 in February. The automobile industry which peaked at 120 point index production in 1937 had dropped to 75.8 in January, 1938.

All these conditions reflected directly upon citrus markets. The quickness with which citrus fruits were dropped from diets in favor of basic foods was indication that citrus was still considered a semiluxury on the grocery list.

Agricultural successes invariably are and have always been dependent upon weather conditions. In this respect, too, Florida's 1937-38 season dealt unkindly with citrus producers. During the mild winter of the previous season, Florida citrus trees had bloomed almost steadily from January through June. As a result, variable maturities for all varieties of citrus during the 1937-38 season presented serious complications in effective grading practices.

A freeze during December, 1937, added to the marketing problems. But prompt action by the Florida Citrus Commission to prohibit the shipment of frost-damaged fruit was successful, for the first time in Florida citrus history, in holding to a minimum the amount of frost-damaged fruit that arrived on the markets. Following the December freeze, the later movement of the Florida citrus crop was seriously affected by a drought more serious than any in previous seasons. Discussion of the drought is recorded in the official minutes of an Exchange board meeting at that time: ". . . it has been so severe and general that fruit has become soft and flabby. Much acreage even under irrigation has been unable to cope with the serious drop of the water table throughout the belt. Trees everywhere are partially defoliated and in some sections are totally so. Younger plantings in some sections not available to irrigation have been killed entirely."

The factions within the industry, including the Exchange, which had pressed for an effective Marketing Agreement, were also dealt what was considered a critical blow during the 1937-38 season. As the annual report of the Exchange for the 1937-38 season indicated:

Apparently unsatisfied with the economic, competitive, and climatic hazards which made the present crop extremely difficult from a marketing standpoint, some Florida operators and growers combined to deny the industry organization and Federal benefits enjoyed by competitive producing sections. California, Arizona, and Texas all have marketing agreements. During part of the last season Florida had a marketing agreement. An analysis of the benefits available from the operation of an instrument of this kind reveals it to be productive and desirable.

Denied the control afforded by a marketing agreement, Florida dumped her 24 million boxes of oranges and 15 million boxes of grapefruit and tangerines into the market. No semblance of orderly marketing existed. This disorganization and chaos directly affected not only Florida's markets but those of other citrus-producing sections as well.

The Florida Citrus Exchange may point with pride to its ceaseless effort for a marketing agreement which would have given the industry protection against such disorganization. Its officers and a committee of directors labored constantly and earnestly from the middle of last summer in an effort to obtain the agreement. In spite of many concessions made to opposing interests, the earnestly sought unity in the industry remained unattainable. The compromise agreement finally offered was rejected by Washington in February. [1938]

Failure to obtain a marketing agreement apparently deprived the industry of the proper machinery through which it might have obtained an effective federal government surplus purchasing program. With the cooperation of other grower-shipper interests, the Florida Citrus Exchange finally obtained the active interest of the Federal Surplus Commodities Corporation in the Florida citrus industry. While the policies of the FSCC were apparently not all that could have been desired, the Exchange felt that the final outcome was of benefit to the industry.

It was in these circumstances that the Exchange again took up the question of some sort of industry organization that could unite growers and shippers into an effective program for the betterment of the industry. A movement that was to become known as the Florida Citrus Producers Trade Association had been joined by the Exchange. The philosophy of the Exchange with regard to the association is explained in the text of a report by General Manager C. C. Commander to the board in June, 1938:

This organization has been hailed by many familiar with Florida citrus difficulties as being the most progressive step since the formation of the Florida Citrus Exchange 29 years ago. Its membership already includes the most sound packing and selling operations in the State, totaling somewhat in excess of 50 per cent of the total crop.

It is probable that owners of 60 to 65 per cent of next season's crop will have been accepted as members by October 1. Membership in the Trade Association is not wide open. It is being limited purposely to those constructive factors representing producer interests which may be expected to work together in harmony for common objectives. Were it not for this restriction, it is probable that 95 per cent of the fruit could be signed up before fall.

The Trade Association forms immediately operable machinery which can be used effectively in correcting such evils in the industry as the multiplicity of containers, the many grades and sizes, the lack of standards in car loadings, and any other practices which in the opinion of 80 per cent of its members deserve attention.

None of these objectionable features exist in the California industry. This is one of the reasons for its continued success. It is believed that harmony of agreement provided by this organization will accomplish the same for the Florida citrus industry.

It is apparent that the stress of economic and marketing conditions which brought about the formation of the Trade Association also made necessary some changes in the operating procedure and sales policies of the Exchange.

Basically, these changes included the shift of more responsibility to the member-associations of the Exchange, the promotion of district brands for the optional use of all packinghouses in the Exchange system, a purification of grades qualifying for the SEALD SWEET label, individual market exploitation of all brands, seasonal pools by varieties, sizes, and grades, the organization of a sales committee from the board, standard containers and carloads, and a further reduction of production costs. While these changes dealt almost exclusively in the interests of fresh fruit, it is interesting to survey also other activities of the season.

The Exchange, realizing the need for the diversion of lower grade and discount sizes of the increasing citrus crop, had concentrated on this problem during the 1937-38 season. This was done in an endeavor to create avenues for such diversion which would accommodate appreciable quantities and which would have definite and measurable effect in relieving the fresh-fruit markets, while at the same time paying a reasonable profit to the grower.

Canning operations in which the Exchange had pioneered as early as 1922 had by this time developed to such an extent that they formed an industry within an industry. Diversions to canneries had grown until they reached as high as 60 per cent of seeded grapefruit by the close of the 1937-38 season. The Exchange, however, felt inclined to develop additional channels through which

Florida citrus might be moved profitably for its producers. Two separate ideas were pursued in an effort to accomplish this end. The first of those ideas was the development by the Exchange of an orange juice vending machine invented and brought to the cooperative by Tracy Acosta of New Smyrna and Godfrey Rockefeller of New York. The machine itself was entirely automatic in operation and was so arranged that the consumer by inserting a five cent piece could observe the complicated workings as the mechanism cut, extracted, and strained the fresh juice into a paper cup.

The second idea involved the perfection of a breakfast food known as SEALD SWEET Breakfast Food, a dry food made from whole oranges. The product was invented by Ramon Bustamante of Cape Girardeau, Missouri, and was brought to the attention of the Exchange by its district manager in St. Louis.

While both of these ideas were to be eventually abandoned at considerable expense to the Exchange, it is apparent that the cooperative was relentlessly pursuing any practical method for improving the circumstances of the Florida citrus growers.

One additional event during the 1937-38 season is worthy of notice. For years the Exchange general offices in Tampa were located in what is now known as the Maas Building. As changes in policy and procedure were implemented by the board, its housing costs had become an important factor in its cost of operation, and therefore of much concern in any discussion by the board of curtailing its charges to Exchange shippers. With this in mind, the Exchange concluded the purchase of an entire building at Florida Avenue and Oak Street in Tampa from the Standard Oil Company at a price which would reduce its annual housing cost by nearly 50 per cent. Removal of the organization to its new headquarters was completed during the summer months following the 1937-38 season, and the organization still maintains this site as its general office.

The annual meeting of the board for the 1937-38 season took place on June 9, 1938. Directors seated at this meeting or subsequently seated included H. G. Putnam of Oak Hill representing the Indian River Citrus Sub-Exchange, W. C. Van Clief of Winter Haven representing Florence Villa Citrus Sub-Exchange, W. M. Moseley of Fort Pierce representing Fort Pierce Citrus Sub-Exchange, L. L. Lowry of Winter Haven representing International

Citrus Sub-Exchange, E. F. DeBusk of Gainesville representing Lake County Citrus Sub-Exchange, A. W. Hurley of Winter Garden representing Lake Apopka Citrus Sub-Exchange, Fred T. Henderson of Winter Haven representing Lake Region Citrus Sub-Exchange, S. A. Whitesell of Largo representing North Pinellas Citrus Sub-Exchange, J. C. Palmer of Windermere representing Orange County Citrus Sub-Exchange, John S. Taylor, Jr., of Largo representing Pinellas Citrus Sub-Exchange, William G. Geier of Windermere representing Plymouth Citrus Sub-Exchange, D. A. Hunt of Lake Wales representing Polk County Citrus Sub-Exchange, C. H. Walker of Bartow representing Scenic Citrus Sub-Exchange, R. J. Kepler, Jr., of DeLand representing St. Johns River Citrus Sub-Exchange, W. W. Raymond of Alva representing West Coast Citrus Sub-Exchange, and H. E. Cornell of Winter Haven representing the Winter Haven Citrus Sub-Exchange.

Officers and department heads elected to serve for the 1938-39 season were W. L. Tilden, president and chairman of the board; A. W. Hurley, first vice-president; C. H. Walker, second vice-president; W. M. Moseley, third vice-president; and L. L. Lowry, fourth vice-president. Following a series of internal conflicts with regard to the board structure of the Exchange, Judge W. L. Tilden resigned from the board on April 21, 1939. Subsequently, C. H. Walker was elected to complete the remainder of the term as president.

C. C. Commander was again elected to the top staff executive position, and the position of sales manager, which had remained open for the prior two seasons, was filled by the election of Fred S. Johnston to that vacancy. L. D. Aulls was named traffic manager; S. L. Looney, treasurer-comptroller; O. M. Felix, secretary; and Counts Johnson, attorney.

CHAPTER 25

1938-1939

FEW GROWERS or handlers of Florida citrus today who were interested in citrus during the 1938-39 season fail to remember the near critical circumstances of that period. Because of the many difficulties confronting the industry during the season and because of the direct influence of these difficulties and their solutions on the development of the Florida citrus industry, this history will, for the moment, depart from its general format up to this point. Without regard either to the chronology of events as they occurred during the 1938-39 season or to the specific events themselves, this chapter will simply attempt to indicate the circumstances as they existed at that time and to present the efforts of the Florida Citrus Exchange to meet these circumstances.

It is important to summarize briefly the general make-up and the trade platform of the Exchange as it existed during the 1938-39 season. It must be remembered that since its inception the Florida Citrus Exchange had been and continued to be owned cooperatively by approximately one-third of the growers in the industry. Its concern in industrial progress, therefore, on behalf of the grower was real and sincere. Its policies and decisions were evaluated and dictated on the simple basis of whether or not these policies or decisions would result in benefit to the growers.

With this in mind, the general program of endeavor of the Exchange at this time can be reconstructed into these five basic areas:

1. Preparing the industry's products to meet consistently uniform standards of grades, thus assuring trade and consumer acceptance and confidence.

2. Equalizing freight charges to make possible a balanced distribution to all markets available to Florida citrus exploitation without carrier penalty.

3. Controlling volume of the crops in line with existing demand at price levels which permit reasonable profit to producers and which coordinate with similar control measures exercised by other citrus-producing sections feeding American markets.

4. Developing by-products and extraordinary channels of trade, plus the expansion of normal fresh-fruit channels to increase the flow of Florida citrus through them.

5. Sponsoring of adequate research to determine additional uses and more convincing sales and advertising materials by means of which Florida citrus demand might be increased.

The season of 1938-39 generally illustrated to a great degree that costly inadequacy continued to exist in Florida's citrus marketing machinery. With the citrus industry almost constantly in the doldrums of low price levels for the past few years, the situation seemed in reality an almost graphic reproduction of the danger signals years ago posted by the Exchange when it warned that industrial organization, permitting the use of ordinary sales fundamentals, would eventually be the only answer to increasing production.

The problems of marketing Florida citrus apparently differed little in the 1938-39 season from those basic problems facing the sale of any other agricultural commodity in volume production. On the basis of the successes as well as the failures in all agricultural commerce, the Exchange felt that it could clearly define the fundamental merchandising requirements for profit. These fundamentals were:

1. Prepare the product uniformly and attractively and with standards of quality consistently maintained in proportion to the price levels sought.

2. Offer only those volumes of merchandise which would yield the desired price levels, and divert to noncompetitive by-products channels or through other methods eliminate surpluses beyond these required volumes.

3. Increase the demand through advertising intelligently prepared and timed to meet sales requirements.

The elements for success in all business which existed during this period, and which most probably are still unchanged today, can be recognized in these fundamentals. The Exchange, in its presentations of the era, felt that any degree of success for the owner-producers of the Florida citrus industry would be directly hinged to a rigid observance of these rules. Beyond that, Exchange philosophy emphasized the opinion of its directors that no one of the three fundamentals was sufficient unto itself; that these fundamentals were closely related and inevitably interdependent. It was undoubtedly this philosophy which had given life and action to most of the Exchange policies, both during the 1938-39 season and those immediately preceding it, with regard to the marketing of Florida's citrus crop. The Exchange believed voluntary co-operation of growers and handlers in obtaining a common objective to be the ideal method through which to accomplish the utmost benefit for citrus growers.

As material proof of an industry that had through the years followed this philosophy to a great degree of success, the Exchange pointed to the California citrus industry. Experience in California, where grower cooperation continued to be far more advanced than in Florida, had demonstrated that federal and state legislation had been almost indispensable in the regulation of its crop for intelligent merchandising. With the knowledge of this situation in California, the Exchange was therefore inclined to keep itself in the leadership of a move for the creation of improved legislation, both federal and state, to secure adequate safeguards to regulate the movement of fruit to its ultimate markets.

It is apparent that, with the entry of the federal government in crop controls in 1933 under the Agriculture Adjustment Act, and the various state laws of 1935 and 1937, the Exchange felt that the first and third of the aforementioned three fundamentals had been fairly well attained. The problem, which during these times was simply stated by the Exchange on numerous occasions, was the failure of the industry to adopt the second of the three fundamentals—the retention of the frequent production excesses of Nature in line with the existent market demand. The economic and marketing advantages of relating supply to demand in order to obtain a correspondingly desirable price was without doubt the goal-simple of the Florida Citrus Exchange in its every representation at the industry level during this era.

Let us now review the aspects of the 1938-39 season itself. There is every indication that the 1938-39 season developed progressively into one of the most disastrous in the entire history of the Florida citrus industry. For example, both seeded and seedless grapefruit brought prices below the cost of production beginning in October, 1938, and, with the exception of a few days in early November, continued throughout the season at below production cost levels. Seeded grapefruit varieties frequently fell so far below production cost that they failed to yield any returns whatsoever on the tree. A similar record of all auction standard-packed orange sales averages for the season indicates that growers also sustained heavy losses in this commodity. Orange difficulties were not, however, so severe nor so continuous as grapefruit losses. They were from all indications somewhat relieved by a relatively steady and profitable increase in prices as the Valencia season got under way.

In retrospect, it is interesting to note that where increases or decreases in Florida orange volumes in the years past had been absorbed by the markets in direct proportion to their capacity to pay, this situation had become altered during the 1938-39 season. It must be observed that during both the 1937-38 season and the 1938-39 season price averages dropped faster than declines in income, regardless of minor crop volume fluctuations. Under previous circumstances, these declines were beyond the ability of the Florida citrus industry to control. But during the two recent seasons, high volume output of citrus had caused citrus consumer prices to dip to a far greater degree than the corresponding dip in consumer income. Thus, while the consumer had money to pay for citrus at reasonable prices, the Florida citrus industry had persisted in so glutting the markets that prices were not held in line with the consumer's ability to pay.

Relative to the production statistics of Florida during the 1938-39 season, it is sufficient to say that as of May 20, 1939, Florida had produced nearly 94,000,000 boxes of citrus for a 25 per cent increase over the 1937-38 season. While California was slightly under its previous extremely high volume season, Texas had registered an approximate gain of 50 per cent.

As Florida contemplated its various predicaments, it must have been clearly apparent that an effective solution of its supply control problems could not be regarded as concerned solely with the establishment of control in any one given citrus-producing area. It would have been useless, logically, for Florida growers to consider methods of control on oranges without some sort of similar action on the part of California. Similarly, in grapefruit, Texas had grown to be almost as large a factor in the grapefruit market as Florida from the standpoint of volume of production. Thus, if Florida voluntarily endeavored to curtail its shipments of grapefruit without parallel curtailment of Texas shipments, the resulting Texas advantage would be obvious. The true significance of the situation seems to underscore the fact that the Exchange was not unsound in its search for effective controls throughout all citrus-producing areas feeding American markets.

With these factors in mind, it is interesting to consider the discussion of the season found in the 1938-39 annual report of the Florida Citrus Exchange:

> It [the season] is a record of distress which can be relieved by organization and cooperation between growers and grower-operators and between producing sections.
>
> It is a marketing situation, however, which has confronted all operators in the industry alike. It will continue to do so. Only those will survive and continue uninterrupted grower service who best are able to get the market for fruit handled at the least cost for the service performed.
>
> From the grower's standpoint, this reduction of service costs between the tree and the consumer has become an imperative economic necessity. He must obtain as big a percentage of the gross delivered return for his crop as is possible if he is even to come close to recovering production costs.

The annual report went on to point out that the Exchange had concentrated its efforts of several years exclusively on the marketing of fresh citrus fruits to bring about such a reduction. It recommended that packinghouse cost reductions were likewise necessary to produce advantageous returns for the grower.

While this season of distress weighed heavily on the Exchange, it continued its dealer service and advertising programs. Its legal department continued to provide acutely needed assistance to members during this period when both business and labor had brought about the enactment of a series of laws requiring the attention of trained legal experts. Likewise its traffic department had continued to render important services to the entire Exchange system. The Growers Loan and Guaranty Company, along with the Exchange Supply Company, maintained normal operations to the benefit of Exchange growers and shippers.

On the export markets a period of activity occurred because of the unsatisfactory Palestine citrus crop, the Spanish civil war, and a late-maturing Brazilian crop. While direct shipments to Europe during the past several years had been considered dangerous from an exchange and credit viewpoint, this situation presented an opportunity for profitable export. The Exchange promptly made arrangements for satisfactory credit and exchange facilities, and began to ship large volumes of Florida citrus onto the European market. While the export returns for the season were not considered excellent, records indicate that frequently the average for export sales exceeded those of the demoralized domestic market of the 1938-39 season.

Amid these circumstances, the board of directors of the Florida Citrus Exchange contemplated the future at its annual meeting on June 8, 1939. To those familiar with the tribulations of cooperative organizations, it need hardly be pointed out that the circumstances of this era made the conduct of the business of the Exchange difficult indeed. The organizational minutes of the annual meeting are filled with volumes of internal recommendations, discussions, and some controversy with regard to corrective policies felt by their authors to be of vital importance to the fresh-fruit business.

To the credit of those in official capacity, however, there is every indication that the seriousness of the 1938-39 season did not prompt a single withdrawal of association or shipper membership, nor did it prompt hasty or fanatic rearrangement of organizational procedure. One prevailing factor apparent in a review of these minutes is that members of the Exchange were generally as well or better paid for their fruit than those operators or growers outside the Exchange system. It may be considered, therefore, that the Exchange looked upon the closing season as a disaster that could hardly be avoided in future years by a general reorganization of sales facilities as they existed at that time. Then, as has always been the case, the Florida citrus industry in major part attributed its problems to a lack of cooperation between its various segments. This analysis was in all probability true, and while the diagnosis earned general agreement, the method of treatment remained evasive.

Seated at the board's annual meeting on June 8, 1939, were H. G. Putnam of Oak Hill representing Indian River Citrus Sub-Exchange, W. C. Van Clief of Winter Haven representing Florence Villa Citrus Sub-Exchange, W. M. Moseley of Fort Pierce representing Fort Pierce Citrus Sub-Exchange, L. L. Lowry of Winter Haven representing International Citrus Sub-Exchange, E. F. De Busk of Gainesville representing Lake County Citrus Sub-Exchange, Fred T. Henderson of Winter Haven representing Lake Region Citrus Sub-Exchange, Charles B. Anderson of Tarpon Springs representing North Pinellas Citrus Sub-Exchange, J. C. Palmer of Windermere representing Orange Citrus Sub-Exchange, John S. Taylor, Jr., of Largo representing Pinellas Citrus Sub-Exchange, William G. Geier of Windermere representing Plymouth Citrus Sub-Exchange, D. A. Hunt of Lake Wales representing Polk Citrus Sub-Exchange, C. H. Walker of Avon Park representing Scenic Citrus Sub-Exchange, Tom B. Stewart of DeLand representing St. Johns River Citrus Sub-Exchange, W. W. Raymond of Alva representing West Coast Citrus Sub-Exchange, H. E. Cornell of Winter Haven representing Winter Haven Citrus Sub-Exchange, and R. J. Kepler, Jr., of DeLand representing Clark Citrus Sub-Exchange.

With the exception of Charles B. Anderson of the North Pinellas Sub-Exchange, who was subsequently replaced by J. W. Smith of Brooksville, the entire board as seated at the June 8 meeting served throughout the following season. C. H. Walker served as president and chairman of the board; H. E. Cornell as first vice-president; W. M. Moseley as second vice-president; J. C. Palmer as third vice-president; and L. L. Lowry as fourth vice-president. There was no change in staff executives for the 1939-40 season.

CHAPTER 26

1939-1940

AS HAD HAPPENED so often in previous years, the 1939-40 season, which had given promise of being a high-volume production season with continued marketing difficulties, was "saved" on the markets by the severe freezes that occurred for five consecutive days during the last two weeks in January, 1940. While the dangerously cold weather caused critical losses to growers in some parts of the Florida citrus belt, most of the Exchange acreage reported a yield of merchantable fruit in volumes sufficient to compensate some of the loss because of the higher market levels enjoyed following the freeze. It is interesting to observe, with regard to the freeze, that most of the increased returns of the period following the cold weather went to cooperative growers and shippers who retained title to their fruit, or to speculators who bought heavily from panicked growers and who were able to market this fruit under market rises that followed the freezes.

Under the circumstances, and not illogically, the heaviest financial losses of the 1940 freezes were incurred by packing and shipping organizations who budgeted their operations and based their costs upon an anticipated volume which proved to be substantially short of the original estimates. Additionally, field and packing labor, supply industries, and carrier systems all lost heavily as a result of the cold weather.

The Florida Citrus Exchange, along with most of the industry, emerged from the freezes with even the most experienced fruit men certain that at best there could be no more than a 50 per cent movement of the total crop on trees at the time of the freeze. Subsequent developments, however, apparently brought to light certain very favorable conditions. Wood damage was not so serious as had been feared, apparently because of the several days of near-freezing weather that preceded the five-day freeze. While defoliation was reported as general, widespread recovery was equally rapid, and weather following the critical period so favored new growth on citrus trees that the blossoming season saw one of the heaviest blooms ever witnessed in Florida.

Production figures for the 1939-40 season, which were originally estimated at 30,500,000 boxes of oranges, produced approximately 75 per cent of that figure as a result of the freeze. Grapefruit production dropped around 15 per cent, with total utilization as of May 11, 1940, estimated at 14,691,904 boxes. Tangerine utilization was also reduced from an early season estimate of 2,900,000 boxes to an actual utilization of approximately 2,300,000 boxes.

At its regular monthly meeting on February 16, 1940, the Florida Citrus Exchange rallied to the assistance of those members who had been placed in financial crisis because of the freeze. The following resolution was unanimously adopted by the board:

WHEREAS, the freeze recently experienced is showing itself to be one of the worst in the history of the industry, creating serious problems among a large number of our grower membership and associations in the proper care and cultivation of their grove properties, and

WHEREAS, these conditions create an unprecedented opportunity for the Florida Citrus Exchange, the Growers Loan and Guaranty Company, and affiliated associations and associate shippers in the cooperative system which they represent, to perform in protecting and assisting their individual members in a manner which will exemplify the advantages of cooperative membership, and

WHEREAS, the declared policy of the Board of Directors embraces the assistance to member associations so that they may obtain sufficient volume to permit their operations to be conducted on an economical and efficient basis, assisting such tonnage acquirement through the Growers Loan and Guaranty Company with such financial cooperation as seemed advisable from time to time, frequently on a short-term basis and from season to season, and

WHEREAS, the Florida Citrus Exchange, the Growers Loan and Guaranty Company, and the cooperative system which they represent are in the best shape in the history of cooperation in Florida, having proved their value over a period of more than thirty years, constantly perfecting methods, sales representation, merchandising facilities, and financial responsibility to the point where the organization stands today as the outstanding operation in the industry, the second largest in the world handling perishables, equipped with the finest, most far-flung salaried and brokerage representation, trade contacts and reputation for dependable performance of any in the field, and operating on a service charge which is the lowest of any responsible operator in the industry today comparing favorably with the cost performance of the California Fruit Growers Exchange in spite of the great disparity between the gross volumes handled by the two organizations, and

WHEREAS, such performance record and current standing make the Florida Citrus Exchange and the Exchange system the most practical method for adoption and use by the industry as a whole in the solution of the marketing problems attendant upon normally increasing volumes, particularly if that performance currently is enlarged to cope with the existing emergency competently and satisfactorily.

NOW, THEREFORE, BE IT RESOLVED that the resources of the Exchange system, including such centralization of borrowing power for qualifying associations and associate shippers, be used to extend such financial assistance as may be required for the proper care of member groves and for the production of a new crop of top quality and volume.

BE IT FURTHER RESOLVED that associations will be assisted to help finance the personal requirements of its members who may be in dire need of assistance, where such loans may appear endorsed by the association and have been passed by its Board of Directors after due consideration of each application on its merits, and

BE IT FURTHER RESOLVED that the qualifying associations be assisted in financing new tonnage on the above terms and conditions, provided the grove, prospective member and association operations are satisfactory to the committee of executive officers of this organization.

The implications of the foregoing resolution are clear. In spite of several years considered less-than-good by the Exchange, including the nation's worst economic depression and high-volume output by the combined American citrus production areas, the freezes of 1940 had found the organization in position to offer extremely good financial assistance to its members for recovery from freeze damages. From this and similar past actions of the Exchange, this history must record that the cooperative has in all its years continued to offer its greatest service in times of adversity and industrial strife.

On the markets, the Exchange faced a trying situation following the freeze. It considered its wholesale and retail contacts vitally necessary in order to acquaint the trade as fully as possible with the dependability of Exchange grade and packs, yet its operational budgets were low. A total of 12 men was used exclusively in this type of work during the remainder of the 1939-40 season. In addition to this "missionary" contact, the normal function of the dealer service crews was continued. Display materials, both SEALD SWEET and Florida Citrus Commission commodity type of materials, were utilized to attract consumers to Florida citrus. Furthermore, the Exchange for the first time entered into a cooperative agreement with many of its district or association brand producers for the development of advertising of these brands in specific markets. These campaigns were financed with funds of which two-thirds were provided by the district or association and one-third by the Florida Citrus Exchange. The advertising campaigns included the use of radio, newspapers, posters, and other media, and were coordinated through the parent organization.

Exchange advertising policy had been redesigned following the Florida Citrus Commission's nationwide commodity advertising program, which had become increasingly effective since its implementation in 1936. While intensive specific market advertising was continued by the Exchange, minutes of board meetings and other records indicate that the organization's nationwide advertising program was allowed to terminate gradually in favor of the Commission's program. The excise tax now paid by Exchange shippers to the Florida Citrus Commission represented to some degree those funds previously spent by the Exchange to conduct its own nationwide program. In addition to its cooperative advertising and its specific markets advertisement, the Exchange also increased its utilization of point-of-sale materials.

A summary of the Florida marketing picture as it was viewed by the Exchange at the close of the 1939-40 season appeared in the annual report of the Exchange for the season:

> By way of summary, therefore, the Florida citrus industry is confronted with increasing volumes which can be offset only by parallel action from all producing areas. Its alternative must be the increase of consumer demand and methods by which that demand may be reached. It lacks organized selling effort.
>
> It numbers among its hundreds of "sales agents" dozens of truck peddlers. It has lost over 50 per cent of its grapefruit tonnage to canneries at prices averaging below production cost and finds that pack advertised and sold in direct competition with the fresh fruit from which it hopes to make a profit.
>
> Only one conclusion may be reached by the logical and unbiased mind in studying these data: a coordination of selling effort between the responsible cooperatives, grower-shippers, and speculative operators will be necessary if the industry is to distribute its constantly increasing volumes at a profit to producers.
>
> Against this background of industrial disorganization and costly self-competition the Florida Citrus Exchange operated with improved service to its grower members through well-managed associations and associate shippers cooperating fully with the Sales Department. This industry situation can be met best with most efficient packing and productive cultural services. This branch of cooperative service to the grower is of paramount importance; his cultural and packing costs represent the greatest percentage, exclusive of freight, in his total handling costs from blossom to market.
>
> Proper administration of these necessary expenditures is of utmost importance as upon them depend production of volumes per acre and improvement in the quality of fruit produced. Better average quality in Florida must be obtained. This is particularly true of the early varieties. Much study to improve both external and internal quality of fruit has been given by many expert horticulturists in Florida. Much progress along this line has been made. All association [Exchange] production departments should confer with Dr. Camp [Florida Experiment Station] and other leaders in this field for facts or procedure which will improve their crops.

It is of interest to note that the Florida Citrus Exchange, during the 1939-40 season, operated on the lowest selling charge in its history. Selling charges based on the high volume expected before the freeze were not increased, owing largely to the establishment some years before of a reserve contingency fund for just such a purpose. Thus, at the close of the 1939-40 season, the Exchange looked forward to gains of more than 6,906 acres of citrus as estimated by its district managers in June, 1940. This acreage, indicated by advance affiliation applications, would represent grove land belonging to nearly 250 new Exchange growers persuaded by Exchange results to cast their lot with the cooperative.

The annual meeting of the 1939-40 board of directors of the Florida Citrus Exchange was held on June 6, 1940. Directors seated at this meeting included H. G. Putnam of Oak Hill, W. C. Van Clief of Winter Haven, W. M. Moseley of Fort Pierce, L. L. Lowry of Winter Haven, J. N. Mowery of Eustis, Fred T. Henderson of Winter Haven, J. W. Smith of Brooksville, J. C. Palmer of Windermere, William G. Geier of Windermere, D. A. Hunt of Lake Wales, Tom B. Stewart of DeLand, W. W. Raymond of Fort Myers, H. E. Cornell of Winter Haven, and R. J. Kepler of DeLand.

Subsequently, the Polk Packing Association successfully applied for membership in the Exchange and was represented on the board by John A. Snively, Jr., C. H. Walker was seated as a representative of Scenic Citrus Sub-Exchange, and W. O. Kirkhuff of Bradenton was seated as the representative of Pinellas Citrus Sub-Exchange.

Officials of the board elected for the 1940-41 season were C. H. Walker, president and chairman of the board; H. E. Cornell, first vice-president; H. G. Putnam, second vice-president; J. C. Palmer, third vice-president; and W. O. Kirkhuff, fourth vice-president.

Department heads elected to serve for the 1940-41 season were C. C. Commander, general manager; Fred S. Johnston, sales manager; L. D. Aulls, traffic manager; S. L. Looney, treasurer-comptroller; O. M. Felix, secretary; and Counts Johnson, attorney.

CHAPTER 27

1940-1941

THE EVER-INCREASING Florida citrus production volume, which had molded a sort of pattern for the future since the mid-1930's, was again apparent during the 1940-41 season. This picture was compounded by a parallel high-volume year in California, Texas, and Arizona. These four major citrus production areas combined to produce an estimated total of 121,-317,000 boxes of citrus fruits, a figure that was 22 per cent higher than the previous five-year average crop. In fact, it was exceeded only during the 1938-39 season when one-half of 1 per cent more fruit was produced.

Florida production itself touched new heights. The orange volume was nearly 20 per cent higher than the five-year average and was equalled only during the 1938-39 season when approximately 8 per cent more oranges had been harvested. Grapefruit production increased over that of any preceding season by more than 40 per cent increase over the five-year average and exceeded the previous record established during the 1938-39 season by some 600,000 boxes. Even tangerine production approached the record proportions of the 1938-39 season and exceeded the five-year average by about 6 per cent.

In addition to all this, indications of new plantings, tree count and maturity in all four citrus-producing sections of the country pointed to a continued increase in the size of the national crop. It was therefore considered more than likely that the ensuing five-year average could be greatly increased over that of the past averages. These factors were sufficiently important to cause Florida growers some concern for future developments if they were to protect their existing investments in the industry.

The 1940-41 season had other problems, none of them new, which continued to plague the Florida industry. Grower organization in the industry still seemed comparatively ineffective. In the words of one Exchange official at the time: "There are even many cooperative growers who unfortunately do not cooperate." As it had done so often in past years, the Exchange publicly deplored the practice of the maintenance of a large number of fresh-fruit marketing organizations within the industry which were in direct competition with themselves and the rest of the industry. It is interesting to note that the 1940-41 season saw fruit sent to market by a total of 613 shippers and growers duly licensed by the Florida Citrus Commission as "handlers." This figure included express shippers and many others who were not packers and canners.

There is indication that some attempt was made to coordinate these marketing activities through the Florida Citrus Producers Trade Association, but the problem was of such nature that the solution continued to evade the voluntary group. The inability of the FCPTA to gain complete cooperation within the industry eventually moved the Exchange temporarily to withdraw its membership from the organization. The Marketing Agreement, although not the same as the one in 1936-37, continued in force, limiting its control to grade and size prorate methods. The Exchange, still seeking volume prorate, was often critical of the efficiency of the Agreement under circumstances existing at that time to meet the pressing distribution problem.

From communications and records still available from this period, it is apparent that the prospect of irregular and uncontrolled movement of high-volume crops was regarded by the Exchange as responsible for a hesitancy on the part of the trade either to make heavy commitments on Florida fruit or to handle it on any basis other than a day-to-day operation. The industry could not, during the 1940-41 season, extend price protection to the trade in its purchases, and buyers were anxious to avoid acquisition and maintenance of stocks that might devaluate by excessive shipments from the state to any given market on the following day.

It was natural, the Exchange reasoned, that many in the trade preferred to place their lot with California citrus during those periods in which it was available in adequate volume. There is little doubt that the better organization of the California industry assured a more uniform movement at this time, and a consequent price protection against the unpredictable peaks and valleys of prices inherent in the Florida citrus movement.

It should be remembered, also, that f.o.b. or direct sales presented somewhat of a problem to the Florida citrus industry. Citrus prices, then as now, were made either on a bid or a barter basis. While auction sales represented the bid method, f.o.b. sales were always consumated by a trade or barter method. And, perhaps much more so at this time than now, prices were established as a result of sales performance at the auctions. It was in this manner that a marketing situation existed during the 1940-41 season which was unfavorable to the free movement and satisfactory pricing of Florida citrus. Deals whereby shippers sold their goods to distributors in those markets served by the major auctions obviously withdrew buying support from the auction and directly created an unfavorable price reaction.

This fact, although well recognized by contemporary citrus handlers, was becoming of major significance with the growth of chain-store operations that felt completely justified in buying citrus direct because of the substantial savings over the handling charges occasioned by the movement of fruit through auction channels. This bypassing of fruit around auction buyers in effect served to restrict patronage at the auction and reflected a low price, forcing a correspondingly low level in the completion of f.o.b. price arrangements. As is nearly always the case in pricing difficulties such as this, it was the Florida citrus grower who became the eventual loser. To make matters worse for Floridians, this marketing problem was to a large extent exclusively peculiar to Florida. Its major competitor, California, with a grower organization strong enough to clear its fruit through any method of sale in any one market, was relatively unaffected by the swing to f.o.b. sales.

A third marketing problem for fresh fruit during the 1940-41 season was the development of satisfactory price levels on fruit moving to canneries. This situation was more alarming than any other to many citrus experts. They based their concern on the fact that fruit was moved to canneries on a salvage low-price basis, frequently at less than the cost of production. Because of disorganization among producers, canners had been in position to take advantage of this situation, passing along their savings to the nation's consumers in direct competition to fresh-fruit trade. The canning problem, as viewed by the Exchange, was not solely one of the 1940-41 season, nor was it limited to grapefruit, which had in the past been the mainstay of the citrus canning industry. By 1940 the total estimated volume of oranges moving into canneries was placed at 15.44 per cent of the entire Florida orange crop.

While Florida's position in the markets during the season of 1940-41 was obviously second best to that of California, and while the state average returns for Florida fruit were generally considered to be less than satisfactory, the Florida Citrus Exchange reported at the close of the season that its sales had brought premium prices for member-growers. These premiums, as listed in the annual report for the season, indicated that the Exchange average against all competition brought premiums ranging from four cents per box on tangerines to twenty-nine cents on seeded grapefruit.

It is significant to observe in this regard that the premiums mentioned in the preceding paragraph were far greater for the Indian River district. The Exchange had, during the season, cooperated with the growers of this district in the promotion of their FLORIGOLD and FLO brands. The district had levied a special assessment on its growers to cover the cost of that campaign. The results were definite, and the continuance of this program until present times has placed the Indian River brands high on the premium list in their distribution areas.

There is indication that international economic conditions as a result of what was soon to become known as World War II were already being felt in the Florida citrus industry. During the early part of the 1940-41 season an opportunity presented itself to the Florida Citrus Exchange. A large percentage of fruit retailers in western Canada were of Chinese origin. As a race they naturally resented the Japanese invasion of their homeland. While their primary source of citrus had been the Japanese Satsuma, the Exchange found that they were waging a sort of economic war against the Japanese even before the Canadian government banned such purchases.

The Exchange promptly developed a tangerine pack similar to the Japanese satsuma pack, and made its first transaction of seventy-five cars at what was considered to be a very favorable price. The shipment was supported with advertising and news articles in the territory to acquaint the trade and consumers with Florida tangerines. This transaction caused a flurry of excitement among Florida tangerine growers and shippers, but Canada's participation in the war soon caused the restriction of foreign exchange and precluded further tangerine sales to that country. The idea of selling tangerines to Canada, however, was planted firmly in Florida, and it was planned that the Exchange would pursue such business as soon as the international situation would permit.

The season 1940-41 was summarized by General Manager C. C. Commander in a report to his board at the close of the season:

> Success or failure in marketing of Florida citrus during any given season depends on several basic variables. During the season 1940-41 the conditions imposed by these factors appeared as a dangerous threat to grower profits.
>
> That the season is being concluded with returns to growers beyond the limits normally outlined by these oppressive marketing factors has not been accidental. The Florida Citrus Exchange in spite of the fact that it must operate with an organized strength limited to about 25 per cent of the industry's tonnage took the lead in the creation and adoption of moves which helped create better markets for growers.
>
> These individual actions of the industry achieved some results. They were hastily conceived and put into action to meet emergencies which never could have arisen under a thoroughly organized grower-controlled industry. Temporary stopgap action of this character cannot be substituted effectively for the fundamental principles governing the successful merchandising of any product. Particularly is that true of Florida citrus, a highly competitive, huge-volume, seasonal perishable.

The annual meeting of the board of directors of the Florida Citrus Exchange was held on June 5, 1941. Directors seated at this meeting to serve during the 1941-42 season, or those subsequently seated, were H. G. Putnam of Oak Hill representing the Indian River Citrus Sub-Exchange, W. C. Van Clief of Winter Haven representing the Florence Special Citrus Sub-Exchange, W. M. Moseley of Fort Pierce representing the Fort Pierce Special Citrus Sub-Exchange, L. L. Lowry of Winter Haven representing the International Special Citrus Sub-Exchange, C. B. Hipson of

Umatilla representing the Lake County Citrus Sub-Exchange, A. R. Updike of Lake Wales representing the Lake Region Citrus Sub-Exchange, J. W. Smith of Brooksville representing the North Pinellas Citrus Sub-Exchange, J. C. Palmer of Windermere representing the Orange County Citrus Sub-Exchange, W. O. Kirkhuff of Bradenton representing the Pinellas Citrus Sub-Exchange, William G. Geier of Windermere representing the Plymouth Special Citrus Sub-Exchange, D. A. Hunt of Lake Wales representing the Polk County Citrus Sub-Exchange, J. A. Snively, Jr., of Winter Haven representing the Polk Packing Special Citrus Sub-Exchange, C. H. Walker of Avon Park representing the Scenic Citrus Sub-Exchange, H. E. Cornell of Winter Haven representing the Winter Haven Citrus Sub-Exchange, and R. J. Kepler of DeLand representing the Clark Special Citrus Sub-Exchange.

C. H. Walker was elected president and chairman of the board for the 1941-42 season, and H. E. Cornell was named first vice-president. H. G. Putnam was elected to serve as second vice-president, and third vice-president was J. C. Palmer. W. O. Kirkhuff was elected fourth vice-president.

Executive positions were appointed, with C. C. Commander as general manager, Fred S. Johnston as sales manager, L. D. Aulls as traffic manager, S. L. Looney as treasurer-comptroller, and Counts Johnson, secretary and general counsel. Added to the executive staff was James Samson of Tampa as assistant secretary. Samson, already a long-time administrative employee of the Exchange, was to assume increasingly important responsibilities for the Exchange and its affiliates. The assumption by Johnson of the secretarial post resulted from the retirement of O. M. Felix after twenty-three years of employment with the Exchange. In this regard, it is interesting to note that Charles Felix, son of O. M. Felix, had at this time been employed by the Exchange for eighteen years, and as this history is published, is completing his thirty-sixth year with the organization.

CHAPTER 28

1941-1942

THE CITRUS SEASON of 1941-42 presents to the historian the most difficult era in the entire scope of this history to edit and record adequately in the truest possible sequence of events and their effect on the Florida citrus industry. It is a matter of record that as the season opened the potential for grower returns was greater than at any time in the history of Florida citrus. Almost every factor affecting prices seemed to be stacked in favor of a most successful period. Economic conditions were at a high peak for American business, and even the high-quality crop volume, by virtue of its shortness, pointed to prosperity for the Florida citrus grower. In the shadow of so great a combination of favorable factors, the Florida citrus industry looked toward the fall of 1941 with considerable optimism.

The industry was soon to change its mind. Unfavorable weather struck almost simultaneously with the beginning of the season. Unseasonably hot weather, heavy rains throughout the state during the late summer and fall, and the combined murky dampness, created ideal conditions for excessive decay in all varieties of fresh fruit. To add to this problem, California for the first time in many years declined to operate under a marketing agreement and the volume control factors made possible through such an agreement. The damage caused by this lack of controlled shipments from California seriously affected pricing problems for Florida, which had heretofore capitalized on the regularity of the West Coast even though unable to provide volume controls for itself.

Thus, Florida entered the season with what appeared to be more than its usual parcel of problems. It was in this predicament that Florida's citrus industry awoke on the morning of December 8, 1941, to find itself along with the rest of the world ensnarled in the complications of international warfare on the largest scale ever experienced by man. There was an immediate reaction in the economic thinking of both the average American and his government. The latter, although still purchasing surplus citrus up to this time under a program mentioned in previous chapters, dropped its purchase program immediately. Nevertheless, the government continued to be the largest potential individual purchaser of Florida citrus fruit. This resulted from the food requirements of the armed forces as well as from this nation's lend-lease arrangements for allied civilian and military populations. These factors placed the government in a far better position to handle surplus citrus than under its prewar program, yet the potentiality needed definite action by Florida citrus before becoming reality.

With price levels continuing to drop during January, it became apparent to the Florida Citrus Exchange that some coordinated program with the federal government was a necessity. The board requested other factors in the industry to assist in developing some plan through which Florida citrus could regain the confidence of the regular trade channels in an effort to improve the plunging price situation. A conference was held by this group with the chairman of the Florida Citrus Commission, Tom B. Swann of Winter Haven. To the credit of Swann and this group of conferees, a plan of action was developed. Out of this plan came the appointment of a representative committee of the industry which was to carry Florida's citrus problems to the nation's capital. This committee included Governor Spessard Holland, long associated with the citrus industry; Swann; Carroll Lindsey, president of the National Canners Association; and C. C. Commander, general manager of the Florida Citrus Exchange.

The members of the governor's committee lost little time in presenting its case to the Florida congressional delegation and to the Department of Agriculture. They reported that Florida citrus was in critical condition as a result of the war-changed economy. The committee met with the British Purchasing Commission, the Quartermaster Corps, and officials in charge of virtually every phase of government food purchases. The results of the committee's activities were almost immediate. Orders were placed for the government purchase of grapefruit, both fresh and in cans, and later purchases of canned orange juice and fresh oranges were effected. Almost overnight the successful action of the committee had a salutary effect on Florida citrus prices. Fears of sugar and tin rationing were dissolved in the rush to handle the government business thus made available. Cannery price levels rose and subsequently fresh-fruit markets strengthened appreciably. While the war years were to produce other difficult problems, the prompt action of the industry in its initial baptism under wartime conditions provided a foundation for its development as the war continued.

That the Exchange was particularly pleased with the action of the Florida Citrus Commission during this period is apparent in the text of a report heard by the board of directors toward the close of the 1941-42 season:

> Actions of the Florida Citrus Commission under the leadership of Thomas B. Swann were helpful throughout the season on many phases of industry action other than the matter of price ceilings and government purchases previously discussed.
>
> The advertising campaign handled by the Commission for the industry has been more productive than any campaign during the Commission's existence. The net worth of advertising on Florida fruit may be judged only by a comparison of Florida price levels with competing sections for the same periods.
>
> The Commission also was helpful in representing the industry in its attempts to eliminate the truck-weight barriers hampering the free movement of Florida citrus to mid-western markets for truck carriers.
>
> As the industry, together with the rest of the country, becomes more involved with rules and regulations, tin and rubber shortages, price ceilings, rationing, wage and hour difficulties, labor shortages, et cetera, the Florida Citrus Commission holds every possibility of being able to represent the industry as a whole successfully. Such efficient representation will be even more important next season than it has been during the current season.

As a whole, the Florida Citrus Exchange could see some advantages inherent in the effects of the war on the industry. Not the least of these advantages would be the increasing rationing of consumer purchases of many other items of foodstuff. The shortage of sugar was already reducing the volumes of bottled synthetics sold through hundreds of thousands of retail outlets across the nation. The refreshment drinking habit, reasoned the Exchange, would offer a marketing opportunity to citrus never before equalled in the history of the industry.

A reference to concentrate and other products in the annual report of the 1941-42 season is interesting in the light of contemporary developments in these fields:

> The shortage of shipping for the transportation of fresh or canned citrus to our Allies created a sharp increase in the demand for concentrated citrus juices, particularly orange concentrate. Two new plants are being rushed to completion in Florida through the aid of Federal financing. These facilities were not available for use in the movement of the current crop but will be a considerable factor next season. In addition to these large plants in Florida, there are seven plants already in operation in California. The continuation and expansion of these facilities for the coming season may do much to divert a substantial percentage of that state's crop away from domestic markets with a correspondingly favorable price condition. A further war-created citrus market was developed this season permitting the manufacture of a marmalade base using both oranges and grapefruit, which can be expanded in direct ratio to available transportation to meet market demands in Great Britain.

The development of new and improved processing methods together with a growing interest in by-products was viewed by the Exchange as a factor in the future of the industry.

The 1941-42 season was, for the most part, considered successful for the Exchange despite setbacks including the withdrawal of the Haines City Citrus Growers Association—charter members in the Exchange system—from the organization. To offset this withdrawal, however, the Exchange enjoyed increased membership during the season as a result of the affiliation of the Great Southern Citrus Association in the Winter Haven area, the Alcoma Citrus Cooperative at Lake Wales, and the Lake Placid Packing Company at Lake Placid. This combined additional tonnage brought nearly a million boxes of citrus into the Exchange system during the 1941-42 season. At the conclusion of the season, the Exchange system included thirty-one grower associations and twelve associate shippers, and had placed on the market a volume of fresh citrus in excess of 25 per cent of the total state shipment for the season.

It will be remembered by those who lived in these times that the close of the 1941-42 season came in the midst of uncertainty for every phase of the Florida citrus industry. The impact of the war economy upon the nation could be expected to have increasing effect upon the economy of the industry itself. Prospects for a heavy production season in 1942-43 had been enhanced by excellent weather before and during the bloom season, and the income level of the nation assured that an almost unprecedented capability existed for the purchase of citrus. Yet Florida looked with concern at the growing materials and labor shortages that would exist in the coming season. The government's requirements for nitrogen seemed certain to cause a scarcity in this plant nutrient considered so vital to the success of cultivation of Florida citrus. Wartime restrictions on transportation had already eliminated boat shipments, and there was every indication that truck transportation would become difficult to acquire by the beginning of the 1942-43 season. The adjustment to wartime government controls was still complicated and the summer months following the 1941-42 season found the citrus industry deeply involved in the application of new and confusing federal statutes to agricultural industry.

As the board of directors met on June 4, 1942, the uncertainty of the course of future developments was apparent in the almost utter lack of business that came before this annual meeting. A resolution which established re-employment of Exchange personnel called into the services during the duration of the war was unanimously adopted, and represented the only important business transacted, with the exception of board elections and reorganization.

C. H. Walker was returned as president and chairman of the board, H. E. Cornell, first vice-president; H. G. Putnam, second vice-president; J. C. Palmer, third vice-president; and W. O. Kirkhuff, fourth vice-president. Seated on the board for the 1942-43 season were R. J. Kepler of DeLand, W. C. Van Clief of Florence Villa, W. M. Moseley of Fort Pierce, H. G. Putnam of Oak Hill, C. B. Hipson of Umatilla, A. R. Updike of Lake Wales, J. W. Smith of Brooksville, J. C. Palmer of Windermere, W. O. Kirkhuff of Bradenton, William G. Geier of Windermere, D. A. Hunt of Lake Wales, C. H. Walker of Avon Park, and H. E. Cornell of Winter Haven.

There was no change in department officials, with staff positions being held by C. C. Commander, general manager; Fred S. Johnston, sales manager; L. D. Aulls, traffic manager; S. L. Looney, treasurer-comptroller; Counts Johnson, secretary and general counsel; and James Samson, assistant secretary.

CHAPTER 29

1942-1945

BY THE BEGINNING of the 1942-43 season, Florida's entire citrus industry was geared to the nation's war effort. Labor, transportation, distribution, sales efforts, cultivation, and every other element of the industry had become limited in freedom of action through government restrictions on all civilian enterprise. Because of this, the history of this period reflects little in the way of variation of policy or activity within the Florida Citrus Exchange or, for that matter, within the entire industry. This history will, therefore, attempt to cover the seasons of 1942-43, 1943-44, and 1944-45 under this single heading. Names of the members of the Exchange board of directors during each of these seasons will be found at the end of this chapter along with the names of the officers and department heads for each season.

It must be noted at the beginning of this review of the war years that the purchasing activity of the government had become the largest single factor affecting the industry by the beginning of the 1942-43 season. All production of canning and processing operations had been requisitioned by the government for military and lend-lease use early in the season. Tin allotments became the basis upon which such control was exercised, and the only exception to this blanket purchase for the 1942-43 season to be noted occurred when slightly over ten million cases of grapefruit juice were made available for civilian purchase and consumption.

This utilization of processed products by the government had a most favorable reaction on the economy of the fresh-fruit phase of the industry. For, in common with other industries producing essential products for military and civilian uses, it enjoyed a seller's market throughout the three-season period. The Florida Citrus Exchange, in this regard, experienced the most outstanding series of seasons in its entire history, demonstrating the values of cooperative service to its members. Fortunately, the Exchange, along with the industry, dealt in products essential to the health of the nation, such classification almost automatically giving it advantages over other industries of less essential nature. Throughout the period the Exchange enjoyed certain considerations in the handling of its products, which tended to ease substantially the normally difficult task of sales and distribution.

The total volume of production of citrus in Florida jumped to new records in the 1942-43 season as the industry reached near the 70,000,000 box mark, and again in the 1943-44 season when production reached 80,800,000 boxes. A hurricane which swept the state on October 19, 1944, destroying an estimated 25,000,000 boxes of fruit, held the season 1944-45 to 69,100,000 boxes total, but even this retarded season reached production levels never attained before the war.

While the commercial interests in the industry struggled through these difficult but prosperous war years, certain developments in citrus research were taking place that were eventually to change the entire future of the industry. During the 1943-44 season Dr. L. G. MacDowell of the Florida Citrus Commission and Dr. A. L. Stahl of the University of Florida Agricultural Experiment Station, working together on the development of a frozen citrus concentrate, were reporting success in their undertaking. This item of interest appears in the Exchange's report for the 1943-44 season:

> How important frozen citrus concentrates become in the distribution of Florida citrus fruit remains to be seen, but it is believed these new products may find a substantial outlet with the institutional trade, with soda fountains, and, in time, through the packaged frozen-food industry. Two plants for the manufacture of frozen citrus concentrates are being equipped at Orlando and the University at Gainesville to perfect the processing and freezing methods. The Florida Citrus Exchange is taking an active interest in this new product, with the view to utilizing it for the benefit of its members if it proves to be practical.

The matter of frozen concentrates is also discussed in the annual report for the 1944-45 season:

Frozen orange juice concentrate received its first extensive commercial test late this season when Peoples Drug Company used it for preparing orange drinks in its Washington, D. C., stores. This product came from the laboratories and was produced in limited quantity for the first time in 1944. Frozen concentrate for the Washington test is being made by Citrus Concentrates, Inc., for Knight and Middleton, Inc., of Dunedin. Preliminary reports indicate that fountain customers like orange juice made from concentrate about as well as that made from fresh fruit; also, because it is easier for them to handle, the drug stores can sell more orange juice when they make it from concentrate. This test shows that thus far the Washington drug firm is selling more boxes of fruit for Florida growers than when it used fresh oranges.

Another interesting note in the same annual report had this to say:

A series of tests with citrus juice crystals or powders and frozen concentrates was conducted by National Research Corporation at a pilot plant at Plymouth, Florida, this season, in cooperation with the Plymouth Citrus Growers Association. These experiments were so successful that plans are being made for the construction of a $1,000,000 plant at Plymouth to produce these products commercially.

The effect of the war on the Florida Citrus Exchange and its affiliated associations, with regard to financial stability, is to this day a favorite topic of conversation. There is no doubt that by the close of the 1943-44 season the Exchange and many of its associations had shed all debt remaining from the depression years. The Exchange itself had accumulated reserves with which to finance future operations, and on every front the financial condition of all Exchange affiliated organizations was considered to be excellent. Prices received by the growers for their fruit during the three-season period enabled many of them likewise to retire indebtedness, and they were also in a position to face the uncertain years ahead with greater financial stability.

While the volume of fruit handled by the Florida Citrus Exchange during this period continued to increase, little of the increased tonnage came from new members. Various communications indicate that every affiliated association in the Exchange system was forced at one time or another to refuse applications for membership because labor and container shortages made it impossible to handle the additional tonnage offered. This situation led to the adoption of a policy by the Exchange that it would serve first those members who owned the facility, refusing to accept new members if by doing so there would be a lessening in any way of the efficiency of its service to its established grower-membership.

Pricewise, there is every indication that Exchange members fared better than their competitors even during the prosperous war years. The following report, heard by the board in May, 1944, is indicative of this opinion: "Until the shipping season ends, in July, accurate comparisons cannot be made of the prices received by Exchange and other growers during the 1943-44 season. But the preliminary reports of affiliated associations show returns to Exchange growers which in nearly every instance are substantially higher than the returns received by most other growers. The differences frequently are 25 cents a box and as high as 50 cents."

But no matter how successful the industry had become during the war years, intelligent citrus leaders throughout the state looked to the future with great concern. This concern was brought into sharp focus during the 1944-45 season when hostilities in Europe were concluded. The future of Florida citrus was largely unpredictable because the economic conditions of readjustments after the war would inevitably affect citrus as well as other agricultural industries.

The steadily increasing production of citrus fruit in Florida, as well as in other producing areas at the time, would most certainly present serious problems in the years to come. Continued plantings of new groves in Florida, which had been extensive during the period from 1942 to 1945 partly because of reduced expense of planting made possible by federal tax regulations and partly because of the prosperity of the recent years, were expected to increase the state's production to well over the 100,000,000-box mark annually. Conversely, the industry could look for a lessened demand for both fresh and processed products during the period of readjustment following the war. In no logical fashion could the industry look to the government for continued purchase of the large quantities of citrus fruit that it had consumed during the war years. In addition, greater supplies of competing fruits could be expected to affect the future distribution and price levels of citrus.

There was also the feeling by many in the industry that changes in supply and demand and in economic conditions would come so quickly that the Florida citrus industry must immediately proceed to find new ways to market more fruit. According to Exchange presentations at the time, one such way would be to have a greatly expanded advertising and promotion program for both fresh and processed products.

In spite of the good conditions which existed at this time, the Exchange had not forsaken its diligent search for effective control of distribution. That this was true is indicated by a report in May, 1944, to the board by General Manager Commander:

> If, in the years ahead, the production of citrus fruit is to be profitable to growers, there must be effective control of the distribution and marketing of our crops, both fresh and processed. Research and advertising, as much as they are needed, will not alone solve the future problems of the industry.
>
> The Florida industry must effectively coordinate the marketing of its fruit so as to be able to work with the California and Texas industries for the cooperation of all producing areas will be needed to find ways to market future crops at profitable prices.

With the conclusion of the war in Europe in May, 1945, the Florida citrus industry could pause to take comprehensive stock of developments during the war period for the first time in three years.

In legislative matters, the Florida legislature of 1945 had enacted a group of twenty bills which had been sponsored by industry groups working in general harmony with each other. The principal effect of the new laws would be to strengthen the Florida Citrus Commission so that the agency would be better equipped to serve the industry in the postwar period. Generally considered, most important of these bills were provisos to change the following:

Qualifications of Commission members to provide that five of the 11 Commissioners would be owners or paid employees of citrus packing, processing, or marketing organizations in addition to holding grower status;

An increase in the Orange Advertising Tax;

A reduction in citrus inspection taxes;

A sizable increase in the state's citrus research program at the Citrus Experiment Station.

The crop outlook at the close of the 1944-45 season was for a short season in 1945-46 as a result of an extended drought which destroyed a heavy early bloom. In May, 1945, the Exchange made the following forecast for the benefit of its members:

> Florida growers should be able to market their fruit at satisfactory prices next season. The outlook for later years is more uncertain. The citrus industry, like most other farming activities, faces a period of readjustment. When the Pacific fighting ends we must find markets for the increase in our production during the war years which government agencies have taken. The problem is greater in Florida than in either Texas or California because our production has increased more.
>
> In 1940-41, before our entry into the war, Florida produced 56,000,000 boxes of fruit. It had a crop of 92,000,000 boxes on the trees this season before the hurricane in October. Much of this increase has been taken up by the government, and it is for that part of our production that we must develop new markets after the war. Production of winter oranges in California has increased none at all during the war years, and grapefruit production in Texas has increased less than in Florida. It will be up to Florida, therefore, to develop a greater postwar civilian demand.

With regard to advertising, Indian River members of the Florida Citrus Exchange had continued the advertising of its FLORIGOLD brand in metropolitan Eastern markets to further development of consumer preferences. Other affiliates of the Exchange resumed their brand advertising efforts in select markets, but there was no concerted advertising effort on the master brands of the Exchange during the 1944-45 season.

The war years had drawn heavily on the talents of the Exchange's legal department and the last year of the war was no exception. Matters of liaison with the Office of Price Administration, the War Labor Board (which had provided war prisoner labor for many growers during the past two seasons), and the Treasury Department, as well as a heavy load of suits against railroads for delayed shipments, had thrown a considerable workload upon this phase of the Exchange organization. By this time the Growers Loan and Guaranty Company was providing financial assistance to meet most of the major needs of all affiliates, and its obligation to the Federal Farm Board had been concluded during the 1943-44 season.

Florida Citrus Exchange

BOARD OF DIRECTORS

SUB-EXCHANGE	1942-43	1943-44	1944-45
Clark	R. J. Kepler	I. J. Pemberton	I. J. Pemberton
Florence Villa	W. C. Van Clief	W. C. Van Clief‡	W. C. Van Clief‡
Fort Pierce	W. M. Moseley	W. M. Moseley	W. M. Moseley
Indian River	H. G. Putnam†	W. S. Buckingham*	W. C. Graves, Jr.*
Lake	C. B. Hipson	J. N. Mowery	J. B. Prevatt
Lake Region	A. R. Updike	A. R. Updike	A. R. Updike
N. Pinellas	J. W. Smith	A. V. Saurman	A. V. Saurman
Orange	J. C. Palmer**	P. C. Peters†	P. C. Peters†
Pinellas	W. O. Kirkhuff*	W. O. Kirkhuff**	W. O. Kirkhuff**
Plymouth	W. G. Geier	B. J. Schwind	A. C. Johnson
Polk County	D. A. Hunt	D. A. Hunt	D. A. Hunt
Scenic	C. H. Walker§	C. H. Walker§	C. H. Walker§
Winter Haven	H. E. Cornell‡	G. B. Aycrigg	G. B. Aycrigg

§ *President and Chairman of the Board*
‡ *First Vice-President*
† *Second Vice-President*
** *Third Vice-President*
* *Fourth Vice-President*

STAFF EXECUTIVES

POSITION	1942-43	1943-44	1944-45
Gen. Manager	C. C. Commander	C. C. Commander	C. C. Commander
Sec. and Legal	Counts Johnson	Counts Johnson	Counts Johnson
Asst. Gen. Mgr.	None	Marvin H. Walker	None
Sales Manager	F. S. Johnston	F. S. Johnston	F. S. Johnston
Asst. Secretary	James Samson	James Samson	James Samson
Traffic Mgr.	L. D. Aulls	E. E. Gaughan	E. E. Gaughan
Treas.-Compt.	S. L. Looney	S. L. Looney	S. L. Looney

CHAPTER 30

1945-1946

DURING the first season following the close of all hostilities of World War II, Florida marketed more fruit than in any other season in the history of the industry. The volume of production substantially exceeded eighty-five million boxes compared with productions of sixty-nine million boxes in 1944-45 and eighty million boxes in 1943-44. While returns per box on some varieties did not reach the heights of the 1944-45 season, the total returns to Florida growers were greater than ever before. That it was Florida's finest season up to this time, there can be little doubt. The price levels received for this highest-production volume are graphic proof that the industry was never in a finer position.

To serve as its board of directors for the 1945-46 season, the Exchange elected E. G. Todd of Avon Park, I. J. Pemberton of Jacksonville, John L. Olson of Dundee, W. C. Van Clief of Winter Haven, W. M. Moseley of Fort Pierce, H. C. Allan of Oak Hill, J. B. Prevatt of Tavares, A. R. Updike of Lake Wales, A. V. Saurman of Clearwater, P. C. Peters of Winter Garden, W. O. Kirkhuff of Bradenton, A. C. Johnson of Mount Dora, D. A. Hunt of Lake Wales, C. H. Walker of Avon Park, and George B. Aycrigg of Winter Haven. C. H. Walker served as president and chairman of the board; W. C. Van Clief as first vice-president; P. C. Peters as second vice-president; W. O. Kirkhuff as third vice-president; and H. C. Allan as fourth vice-president.

There was no change in the appointment of staff executives for the 1945-46 season. They were C. C. Commander, general manager; Fred S. Johnston, sales manager; E. E. Gaughan, traffic manager; S. L. Looney, treasurer-comptroller; Counts Johnson, secretary and general counsel; and James Samson, assistant secretary.

The Exchange and many others had been apprehensive of the postwar possibilities for Florida citrus—an apprehension supported by both government and industry economists at the time. This fear was based on the fact that the government, which purchased fourteen million boxes of Florida citrus during the 1944-45 season, would purchase practically no citrus with the close of the war. Official forecasts were that unemployment in the nation could be expected to reach eleven million people in the first year of the readjustment period. While it is true that the government purchases of citrus were but a small portion of its volume during the war years, most of the apprehension of the past year failed to materialize as expected. The greater supply of citrus for civilian distribution was readily taken up by the consumers of the nation, unemployment did not exceed 2,500,000 at any time during the season, and the supply of competing fruits became only slightly larger than in the preceding season.

Another important factor affecting Florida orange prices was the small production and small sizes of the California Valencia crop during both 1944 and 1945. These conditions greatly boosted the consumer demand for Florida oranges in both the opening and closing months of the 1945-46 season.

Price ceilings, imposed by the war effort, were suspended from November 19, 1945, to January 4, 1946, but were reinstated after orange and tangerine prices climbed quickly because of light supplies and abnormal Thanksgiving and Christmas demand. Maximum prices of canned citrus products were removed early in the season, but prices paid for oranges by canners apparently did not exceed former ceiling reflections until the last few months of the season and then went no higher than under the ceilings of the 1944-45 season.

It is obvious that the greatest single factor affecting the record returns for Florida growers during the 1945-46 season was that consumers, with more money than ever before, were eating more and better foodstuffs. Records of utilization of this era seem to point out that most of the increased consumption of citrus fruits during this period occurred with families formerly in the lower income brackets. At any rate, it was the opinion of the Exchange, as well as that of other competent food-trend observers, that citrus would remain on the tables of this category of purchasers.

There were encouraging developments on almost every front of the citrus industry during the 1945-46 season. The first successful commercial canning of tangerine juice had been reported, with a utilization of 500,000 boxes of tangerines of grades and sizes which could not otherwise have been marketed. Consumer acceptance of this product was reported as good, and Floridians looked forward to a greatly expanded distribution of the "zipper-skinned" fruit.

The grapefruit canning interests, which had sold virtually all their products to the government during the war, were somewhat handicapped by lack of sectionizing labor, giving rise to research into the possibility of mechanical sectionizers. Various reports with reference to grapefruit sections indicate that the trade demand far exceeded the supply packed during the 1945-46 season. This situation induced numerous canners, who had heretofore packed only citrus juices, to build and equip sectionizing plants for future utilization of grapefruit production. Development of new packs of citrus-fruit salads was also under discussion throughout the processing industry.

Distribution of frozen concentrated orange juice, started in the previous season, continued to expand during the 1945-46 season. The distribution of this product was expected to exceed several hundred thousand gallons by the close of the season, compared to a distribution of around fifty thousand gallons during the 1944-45 season. Even greater increase in this distribution was expected as more processors and distributors entered the field. By the close of the 1945-46 season the Plymouth plant of Vacuum Foods Corporation had commenced packing frozen orange juice concentrate and was also expected to produce dehydrated or powdered orange juice by the 1946-47 season.

In the field of research, one notable development was the discovery that thiourea would control stem-end decay, which had always been one of the most serious problems in the distribution of Florida's fresh fruit. Test shipments already conducted by the government had proved that the thiourea treatment was extremely successful in the control of this decay. While permission to use the substance commercially for this purpose had not yet been granted by the United States Food and Drug Administration, the industry was optimistic with regard to the elimination of an old enemy.

Also, relative to scientific research during the 1945-46 season, another problem was being tackled. At the State Cattle Experiment Station in Hardee County a three-month test had been conducted to establish the value of feeding waste citrus fruit to range cattle in the winter months when pasture feeding was limited. In this test whole grapefruit were fed to cattle, with detailed checks on the poundage consumed and weight gains of the cattle. Although details of the test had not been revealed at the close of the season, it was reported that the cattle were thriving on their grapefruit diet. The value of this by-product utilization was of interest to the entire industry, and the tests were no doubt the foundation of the large citrus pulp industry as it is known today.

Other developments of the season which looked favorable during this period were the expected reopening of the British markets to the United States, the shipment of more than a hundred thousand boxes of fresh oranges to Sweden, smaller shipments of both fresh and processed products to Holland and Belgium, and the potential export market outlook in the Scandinavian countries.

The Florida Citrus Exchange fared well, of course, during the 1945-46 season. Perhaps its greatest single difficulty was in the marketing of the great multiplicity of late blooms of the season. One Exchange estimate placed a figure of 30 per cent of all fruit for the season from blooms that appeared after April 1, 1945. The later maturity of this fruit delayed harvesting but tended to stabilize early season prices at high levels. More oranges and grapefruit were processed during the 1945-46 season than in any other year on record. Orange processing leaped upward to nearly 36 per cent of total production, and grapefruit processing continued at around 69 per cent of total production.

While optimism abounded throughout the length and breadth of the Florida citrus belt at the close of the 1945-46 season, the Florida Citrus Exchange took stock of problems that would be faced in the next season. One such problem was the matter of freight rates. Railroads had asked the Interstate Commerce Commission for permission to raise rates on citrus fruits by 15 cents per hundred pounds, while the War Shipping Administration had requested an increase of twenty cents per box on rail rates of Florida fruit bound to North Atlantic ports. Hearings on these rates were already under way at the close of the season.

The canning industry had been advised that the Food and Drug Administration would conduct hearings at some time in the near future to establish standards of identity and quality for canned citrus products. With regard to this impending revision of grades and establishment of standards, the Exchange's position was outlined in a report by the general manager to the board in May of 1946:

The experiences of the 1945-46 season have demonstrated the need for practical, adequate standards of quality to make for greater uniformity in canned citrus products. With the mixing by canners of fruit of different varieties, and from different blooms, there has been a wide range of quality in canned citrus products—especially blended orange and grapefruit juice. To meet the future competition of other fruit juices, it is essential that there be standards of quality for citrus products which will meet the favor of the consuming public. This is one of the most complex, as well as one of the most important, problems of the Florida citrus industry, affecting not only canners but growers, too.

There is indication that the use of eight-pound bags for oranges had been continued by the Exchange and had increased in direct proportion to the available bags. Other plans for consumer packages of citrus fruits were apparently delayed because of the material shortage as well as the shortage of transportation. It is evident, however, that the Exchange recognized a trend toward consumer packaging of citrus fruits and had already asked large retail outlets to study the situation.

It is interesting to note, in the annual report for the 1945-46 season, this item with regard to cost of fresh versus processed fruit:

Average tree-to-car costs of handling fresh Florida oranges and grapefruit increased almost 60 per cent during the period 1938-39 to 1944-45. In the same period, the cost of processing citrus fruit showed a smaller increase, as less labor is involved in canning.

All changes in cost factors are favoring the distribution of processed fruit to the disadvantage of fresh fruit. The prospective increase in freight rates, higher wood container costs, and higher labor costs will further tend to make it more economical for consumers to obtain their citrus fruits in processed form.

Fresh fruit markets must be maintained, because Florida growers cannot depend upon the processing industry to utilize all of their crops at profitable prices. The widening differential between fresh and processed distribution costs makes essential every possible improvement, every possible economy, in fresh-fruit packing, distribution, and merchandising, to meet the increasing competition of processed products.

It is quite possible that the growing concern over the competition of processed citrus products was responsible for the actions of the board on May 17, 1946, when it authorized the re-establishment of a department within the Exchange to handle the sale of processed products of affiliated associations and associate shippers operating processing plants. The action followed a rather lengthy discussion and a comprehensive survey of the problems of marketing processed fruit. Under the new Exchange canning program, the use of the Exchange trademarks would be limited to participating plants, and standards of quality would be established for products packed under these brands. The use of the master brand SEALD SWEET would be limited to canned citrus products meeting the minimum standards of the "fancy" grade.

Thus the board completed another cycle in the change of policy through the years. It will be remembered that the Exchange had pioneered the entry of Florida into citrus processing in the 1920's, and had actually financed the construction of various plants for the purpose of canning grapefruit "hearts." The marketing of processed grapefruit products had been undertaken at the same time. This interest in processing had continued until the disastrous times of the big depression when the board, faced with retrenching necessity, had narrowed Exchange activity to the exclusive pursuit of fresh-fruit sales.

But the marketing of canned citrus products was again to be taken up by the Florida Citrus Exchange, which would handle these products cooperatively in accordance with state and federal cooperative marketing laws. An advisory committee would be established consisting of one representative from each processing plant controlled by affiliates of the Exchange, and a representative of the Exchange, to advise the board of directors in all matters relating to its canned citrus marketing.

The 1945-46 season closed out later than usual because of the bloom condition mentioned earlier in this chapter. By May 4,

1945, however, it was reported that seasonal auction prices of Florida oranges would average $4.51 per box, while grapefruit prices would average $3.87. Perhaps the most spectacular weekly average for Florida citrus went to tangerines, which reached $7.00 per box during the period from November 19 to January 4. Total volume of sales of the Florida Citrus Exchange during the 1945-46 season was placed at $65,000,000—the highest in its thirty-one years of cooperative marketing. The percentage of interstate shipments of fresh Florida fruit handled by the Exchange had increased substantially during the season, and from all points of view the season was the most successful year in Exchange history up to this time.

The annual meeting for the 1945-46 season was held on June 6, 1946, with no change made in the board for the first time in the history of the cooperative. Officials of the board also remained unchanged, and with the exception of the addition of Carlisle Kyle to fill the newly reactivated post of advertising manager, staff positions were all carried over with the personnel of the 1945-46 season.

CHAPTER 31

1946-1947

THE 1946-47 season had no sooner gotten under way when difficulties and setbacks crowded in to arouse the industry rudely from its gold-spun recollections of the 1945-46 season. The period opened with the government estimate of 101,-700,000 boxes of citrus for the season. Concurrently, the Florida Citrus Commission released reports showing stocks of canned citrus juices to be about 17,000,000 cases, an unprecedented carry-over by at least 10,000,000 cases. Since Florida was packing an estimated 75 per cent of all canned citrus juices in the country, these figures indicated that the industry had actually sold to consumers but 78,500,000 boxes of its 1945-46 production of 86,000,000 boxes. It would begin the 1946-47 season with a new crop plus this carry-over inventory for a combined total of 109,200,000 boxes, or 30,-700,000 boxes more than had been sold to consumers in any previous season.

Under these conditions, it was a matter of routine that Florida citrus prices declined under heavy shipments and nationwide publicity about overproduction and surplus stocks shortly after the opening of the season. By early December most orange sales were returning less than what the Exchange considered to be average production costs, and by mid-January grapefruit and tangerine sales were at comparably low levels. By early February prices had descended to critically low levels. Canners were paying scarcely enough for oranges and tangerines to cover the cost of harvesting and hauling, and just slightly more for juice-type grapefruit. Then, as has happened so often during the history of Florida citrus, Nature took control of the situation. Unusually warm weather late in the period from September through January preceded a February freeze by only a few days. The freeze curtailed production by an estimated 20,000,000 boxes to a total of 87,000,000 boxes.

After the freeze a seven-day shipping embargo imposed by the Florida Citrus Commission restored fresh orange and grapefruit prices to fairly profitable levels, and the psychological effect of the freeze losses served for a time to hold a profitable price level. But as shipments resumed in heavy volume, fresh-fruit prices gradually declined again. By mid-May prices were below levels that would return profits to growers.

While the profits-picture was anything but favorable during the 1946-47 season, certain noteworthy developments were occurring with regard to distribution of Florida citrus. With an eventual harvest of approximately the same proportion as that of the preceding season, sales to consumers in both fresh and processed form were greater to the extent that they utilized the season's harvested production. By May 17, 1946, Florida fresh-fruit shipments were about 1,500,000 boxes greater than for the 1945-46 season, with indications that they would reach more than a 2,000,000-box increase by the conclusion of the season. Reported increases in retail sales of canned citrus juices were expected to take an additional 6,000,000 boxes. It seems reasonable to assume that lower retail prices were largely responsible for this situation.

An analysis of the factors affecting fruit prices throughout the 1946-47 season appears in the annual report of the Florida Citrus Exchange for that season:

A combination of unfavorable circumstances were responsible for the lower prices received by the Florida growers this season. They were the first signs of some of the problems faced by the industry in its economic adjustment under postwar conditions.

Supplies increased, not only of citrus fruits but of competing fruits. The production of winter oranges in all states reached 53,460,000 boxes compared with 46,860,000 boxes last season. Apple production, while normal, was the highest in three years. Bananas were far more plentiful.

Demand continued at a high rate, but within price limitations, because increasing living costs left consumers with less money for food stuffs. Consumer resistance to high prices became more pronounced. The housewife became more selective, in quality as well as in price.

Government purchases, which took 20 per cent of all United States citrus crops in the war years, dropped precipitously. In Florida the government bought but small amounts of citrus juices for its armed forces and of orange juice concentrate for the school lunch program.

Quality was also an important factor. The unusually hot weather in fall and early winter months affected the flavor, appearance, and condition of Florida crops. California commanded much higher prices principally because our fruit did not meet trade acceptance.

The quality of much of the carry-over stocks of canned citrus juices, packed from late-bloom 1945-46 crops, was unsatisfactory. It was the quality of these old stocks, as much as the quantity carried over, that caused price declines at the opening of the 1946-47 season.

High prices charged by many retailers for fresh Florida fruit, months after early f.o.b. prices declined, slowed consumer sales and added to marketing problems. The large retail groups which keep their prices in line with costs were of great help.

Price competition of Texas was a serious problem in marketing Florida grapefruit.

The decline in crop values after the opening of the season was hastened by the trade's refusal to buy more than its current needs of new pack citrus juices, in expectation that prices would go still lower. Because of the large carry-over, its current needs were small.

It was not until after the first of the year, when wholesale and retail canned juice prices reached ridiculously low levels, that trade purchases increased. But even then, and throughout the season, trade factors were reluctant to buy juices extensively for future sale.

The efforts of the government to bring about a reduction of all prices was particularly harmful to the marketing of late Valencia oranges. The psychological situation which it created among consumers and retailers tended to force Florida fruit prices, already low, to even lower levels.

The Exchange progressed very little during the 1946-47 season in the development of future markets, although some progress could be seen in the development of beverage bases from citrus.

In the foreign-trade field, the Exchange was watching closely the results of a comprehensive study of world market opportunities being conducted by the Florida Citrus Commission. The Florida Citrus Producers Trade Association, long since reactivated and rejoined by the Exchange, had obtained a British offer for the purchase of 1,000,000 boxes of fresh Valencia oranges during the season, but the offer apparently occurred during the slight domestic market advance following the freeze and was not accepted.

The Exchange's relatively new canning division, hampered by the expected slow-moving initial progress in production, had been successful in obtaining nearly a hundred brokers by the close of the season, but had made distribution to seventeen states and three Canadian provinces in only limited quantities.

Activities of the Exchange's advertising department were increased substantially during the season, partially in expectation of the increased advertising activity for those processed products handled by the organization. It is interesting to note that equal emphasis was placed on both fresh and processed products wherever SEALD SWEET promotions were conducted.

The Indian River district, by now a veteran advertiser of its FLORIGOLD and FLO brands, continued its aggressive campaign in its specific distribution areas, and continued to command premium prices for its brands.

Other associations, preparing to identify their fresh fruit by stamping methods, were planning considerably more advertising in the season to come, and a total of seventy thousand pieces of point-of-purchase material was distributed by Exchange dealer service crews at work in various market areas.

On April 30, 1947, the Growers Loan and Guaranty Company completed its thirtieth year as an affiliated agricultural credit corporation of the Florida Citrus Exchange. As usual, the difficulties of the 1946-47 season saw the company embarked on a full program of assistance to Exchange growers and affiliates. Happily, the war years had placed this arm of the organization in sound financial condition, with ample resources to finance all sound loan requirements of the associations and their members.

The Exchange Supply Company concluded its thirty-first year of service during the 1946-47 season. Conditions during the season, with regard to supplies and materials, had eased somewhat from the difficulties of the war years, and by the close of the season the Exchange Supply Company could assure all Exchange members of prompt service during the following season.

The annual meeting of the board for the 1946-47 season took place on June 5, 1947, with the following directors seated to serve during the 1947-48 season:

E. G. Todd of Avon Park representing the Avon Park Citrus Sub-Exchange, I. J. Pemberton of Jacksonville representing the Clark Special Citrus Sub-Exchange, John L. Olson of Dundee representing the Dundee Sub-Exchange, W. C. Van Clief of Winter Haven representing Florence Villa Citrus Sub-Exchange, W. M. Moseley of Fort Pierce representing the Fort Pierce Citrus Sub-Exchange, H. C. Allan of Oak Hill representing the Indian River Citrus Sub-Exchange, J. B. Prevatt of Tavares representing Lake County Sub-Exchange, A. V. Saurman of Clearwater representing North Pinellas Sub-Exchange, F. W. Moody of Palm Harbor representing the Pinellas Sub-Exchange, P. C. Peters of Winter Garden representing Orange County Sub-Exchange, Armer C. Johnson of Mount Dora representing the Plymouth Sub-Exchange, D. A. Hunt of Lake Wales representing the Polk County Sub-Exchange, C. H. Walker of Avon Park representing Scenic Citrus Sub-Exchange, George B. Aycrigg of Winter Haven representing the Winter Haven Sub-Exchange, and A. R. Updike of Lake Wales representing the Lake Region Sub-Exchange.

Officials elected for the 1947-48 season were C. H. Walker, president and chairman of the board; W. C. Van Clief, first vice-president; P. C. Peters, second vice-president; A. V. Saurman, third vice-president; H. C. Allan, fourth vice-president; Counts Johnson, secretary and legal counsel; James Samson, assistant secretary; and S. L. Looney, treasurer. C. H. Walker, elected president at this meeting, was destined to serve only a short time as the head of the cooperative. His death on October 12, 1947, created a vacancy that was filled by W. C. Van Clief, first vice-president. This move caused an eventual realignment of the officials of the board, resulting in the advancement of P. C. Peters to the office of first vice-president, A. V. Saurman to second vice-president, H. C. Allan to third vice-president, and John L. Olson to fourth vice-president.

Department heads elected to serve for the 1947-48 season included C. C. Commander, general manager; Fred S. Johnston, sales manager; S. L. Looney, comptroller; Carlisle Kyle, advertising manager; E. E. Gaughan, traffic manager; and C. A. Seehof, canning division manager.

CHAPTER 32

1947-1948

WHILE THE NATION as a whole continued to enjoy one of the greatest and longest periods of uninterrupted prosperity in its history, the Florida citrus industry seemed incapable of elevating its own economy during the 1947-48 season. This situation was viewed by the Florida Citrus Exchange as being both exasperating and needless in view of the various marketing factors that existed throughout the season. Considering that the 90,900,000 boxes of Florida citrus produced during the 1947-48 season had been the state's largest crop to date and that it had been produced within a period of higher costs and services, it seemed apparent to the Exchange that disorderly marketing had once again taken its toll from the Florida citrus grower. The size of the crop had been affected by a freeze and two hurricanes California citrus had also been curtailed because of freezing weather on the West Coast, so the total American production of citrus was considerably under its top potential. This circumstance, together with the high national income, would normally have resulted in favorable returns to the grower. That it did not, reasoned the Exchange, was a monumental warning of more difficult times to come as production reached toward the level of a hundred million boxes.

Generally, Florida growers were receiving less than their cost of production during a period of good times, and curtailed production was reason enough for the industry to consider what could be expected when full production potential was reached, perhaps at a time when the national economy would be less favorable to it.

Although the crop was larger in the 1947-48 season, the state had shipped less fresh fruit to market by April 24 than it had on that date the preceding season. Canneries, however, had used considerably more oranges and slightly more grapefruit. Despite this drop in total shipments, accompanied by a similar drop in both California and Texas shipments, Florida oranges and grapefruit were averaging from twenty-five to thirty-five cents less per box at auction than in the 1946-47 season during the last two weeks of April.

A report to the board by Sales Manager Fred S. Johnston during this period is indicative of Exchange thinking at the time:

Again this season, Florida has received less money per box for its fruit than California, despite the fact that Florida's is one fifth of a bushel larger, and a box of Florida oranges contains from five and one-half to six gallons of juice, compared with only three and one-half to three and three-fourths gallons for a box of California Navels. Considering that Florida offers the customer more, and receives less, there must be something wrong with the way it's being offered.

Once again there could be little doubt that Florida's failure to arrive at some sort of cooperative marketing distribution plan had been effective in making a poor season out of what logically should have been a good one. However, the times of adversity were combining to set the scene for another attempt at cooperation by the industry. The following article appeared in the Exchange's annual report for the season:

There was considerable comment throughout the state this season on the obvious need for cooperative marketing in the Florida citrus industry. It was pointed out that the industry needs an organization that would control a big majority of the tonnage and lend its influence to the plan to bring canners under the Federal Marketing Agreement Act, and work toward a state volume prorate.

The Florida Citrus Mutual, a proposed organization that would permit participation by growers, canners, and shippers, is being considered by the industry, and it is hoped the industry will support this organization to cure the existing marketing evils.

The Exchange, through the action of its board of directors, went on record as favoring any cooperative plan that is legal, practical, fair, and for the good of the growers.

The official minutes of the Florida Citrus Exchange show this action taken by the board on February 26, 1948:

A RESOLUTION

BE IT RESOLVED BY THE BOARD OF DIRECTORS OF THE FLORIDA CITRUS EXCHANGE IN REGULAR MEETING DULY ASSEMBLED:

Section 1. That the Florida Citrus Exchange approves the principles of cooperative marketing as will be exemplified by and through the operation of Florida Citrus Mutual.

Section 2. That the Florida Citrus Exchange shall become a member of Florida Citrus Mutual, and the proper officers of said Exchange shall execute the membership contract required by said Mutual, the same to be effective when similar contracts, representing 75 per cent of the citrus fruit production in Florida based on the 1946-47 season's production, have been executed.

Section 3. That in order to assure the fullest measure of benefit to the affiliated organizations of the Florida Citrus Exchange through the latter's membership in said Mutual, said affiliates of the Exchange shall execute the standard grower contract prescribed by the Mutual.

Section 4. That this resolution shall take effect immediately upon its passage and adoption.

An additional item in the minutes of the February 26, 1948, meeting of the board provided that the executive officers of the Florida Citrus Exchange were authorized to make contributions on behalf of the Exchange to the organization expense of Florida Citrus Mutual. From communications and other records in the Exchange files there is little doubt that the movement for the organization of Florida Citrus Mutual gained the immediate favor and support of the Exchange. While the individual services of Exchange personnel in this respect are not officially recorded, this also was most certainly considerable.

With regard to the 1947-48 season, it is interesting to note that more than one-half of Florida's crop was utilized by canneries. Sales of canned orange juice were up more than 40 per cent over the previous twelve-month period and, as a result, canneries utilized more oranges than ever before. Meantime, although the overwhelming majority of the grapefruit crop was going into cans, the total grapefruit pack lagged behind the greatly increased orange pack. It seems certain that sale of grapefruit juice was being affected at this time by the rather poor-quality grapefruit pack during the war years.

The federal government, during the 1947-48 season, purchased an estimated $2,918,000 in Florida citrus products, mostly single-strength canned juices, and concentrates for use in school lunch and foreign-aid programs. Concentrated orange juice was being used increasingly in the school lunch program. A purchase in September by the government of 78,036 gallons was followed later in the season by a purchase of 672,000 gallons. Prices ranged from $2.40 to $2.55 per gallon for these purchases.

The frozen concentrate plants at Plymouth, Lake Wales, and Dunedin had operated throughout the 1947-48 season at near capacity. Indications are that the production of these three plants was of sufficiently high quality to meet with favorable consumer acceptance wherever introduced. This led to the general belief in the industry that citrus products could be marketed profitably when buyers had confidence in the quality of the product. Unfortunately, the industry had established no control of the quality of single-strength canned juices, and its control of the quality of fresh fruit was not effective to a large extent. The Florida Citrus Exchange, realizing this necessity for better standards for fresh fruit, appointed during the season a permanent grade committee comprised of experienced packinghouse managers to recommend standards of quality for SEALD SWEET citrus. This was an effort to combat the effects of the relatively poor quality of the 1947-48 season caused by excessive rainfall, hurricanes, and freezes.

On another front in the industry the Exchange had established a $1,000 scholarship in cooperative marketing to be awarded annually to an outstanding graduate of the College of Agriculture at the University of Florida. Other research projects of the season included important developments at Florida Southern College's new citrus school in the study of vitamin P and the continued search for practical methods of selling citrus juices in the beverage field.

One phase of Florida's citrus trade, that of the export market, had been completely eliminated during World War II. That this was an important economic factor for Florida, particularly during the 1947-48 season, is obvious in the light of the United States prewar fresh-citrus exports, which had totalled ten million boxes of fresh citrus and over five million cases of canned citrus products.

Although the Exchange had planned to re-enter the European and Asiatic trade markets during the 1947-48 season, a dollar shortage in those countries forced the organization, as well as all of the Florida exporters, to postpone the development of these potentially important avenues established by the Canadian government on November 18, 1947, in an effort to conserve the dollar exchange. This quota system, which amounted to an embargo insofar as Florida growers were concerned, resulted in the Canadian purchase of Mexican and Italian citrus, sometimes with American dollars, to the detriment of the Florida citrus industry.

Probing the export market potentiality at the close of the 1947-48 season, the Exchange believed that export markets existed for fresh fruit in nine European countries and for canned citrus fruits and juices in twenty foreign countries. Following the close of the season a representative of the Exchange toured Europe to make further studies of the export possibilities.

The board of directors of the Exchange met on June 3, 1948. The past season stirred no fond recollections for the Exchange as it convened in annual meeting, and the outlook was that the coming season would bring its own parcel of problems. Indications were that Florida's 1948-49 citrus crop would be even larger than the record-breaking production of ninety-one million boxes for the 1947-48 season, and weather conditions had been favorable up to this time for a better quality crop.

Buyers were becoming more cautious and were, as a rule, demanding better quality in citrus as well as in all other food products. Aware of this fact, the Florida Citrus Commission had appointed a grades committee to work with the state and federal departments of agriculture in an endeavor to establish better standards for Florida fruit. The Exchange looked forward to some improvement in the standards for fresh fruit by the beginning of the 1948-49 season. At the same time an effort was also being made to improve the quality of canned citrus juices, but there is some indication that progress was not expected in this regard during the coming season.

Florida could expect keen competition from Texas grapefruit in the new season, and California groves would, because of drought during the spring, produce less than normal crops. This could result in higher prices for Florida producers if other factors were held in line.

All citrus-producing states were apparently cooperating in efforts to sell citrus in large quantities through the European Recovery Program, but the extent to which it would be included in the program would largely depend upon the willingness of the sixteen participating countries to accept citrus. Unquestionably, the volume of citrus utilized through the four-year Marshall Plan would be decided by those countries involved, along with our government, in relation to the availability, price, and keeping qualities.

The United States market most certainly needed further development in the meantime by more intensive efforts to sell citrus on the basis of its health benefits, through the promotion of citrus salads and desserts, and by the production of better-quality fruit.

In summarizing the 1947-48 season, the Exchange could do so with the knowledge that it had sacrificed a large and expensively produced crop during a period of national prosperity, because of the failure of the industry to cooperate on an effective, orderly marketing program designed to protect the dealers and consumers by preventing the shipment of poor-quality, cheap citrus products. Florida's citrus growers were faced with the necessity of regaining the confidence of both consumers and dealers in the matter of quality products, and through some method they would need to eliminate the competition of poor-quality canned citrus.

Meanwhile, the canning division of the Exchange, now in its second year, had been creating excellent consumer demand for SEALD SWEET products. While prices remained at less than satisfactory levels throughout the season, it was generally felt that top-quality and top-grade products would bring improved prices shortly after the close of the fresh-fruit season.

The Exchange's advertising program, in support of the SEALD SWEET trade-mark for both fresh and canned citrus, was extensively pursued during the 1947-48 season. Newspaper advertisements appeared in 43 daily papers twice a week in 22 principal markets. Local retailers in many markets supplemented Exchange advertising by including SEALD SWEET in their own advertisements. The Indian River district again enlarged its advertising program that had, for a number of years, been instrumental in obtaining a premium price for its brands.

Other departments within the Exchange system reported extremely heavy activities during the 1947-48 season. Both the legal and traffic departments carried heavy workloads normally associated with high-volume seasons, but the Growers Loan and Guaranty Company felt the full brunt of two successive seasons of low fruit prices.

The Exchange Supply Company, in a move to appeal for more support from each of the associations in the Exchange system, had been in the throes of reorganization during much of the 1947-48 season, but had nonetheless continued to serve the organization as it had done in the past.

Directors and their sub-exchange affiliations for the 1948-49 season were E. G. Todd of Avon Park representing Avon Park Sub-Exchange, I. J. Pemberton of Jacksonville representing the Clark Sub-Exchange, John L. Olson of Dundee representing Dundee Sub-Exchange, W. C. Van Clief of Winter Haven representing Florence Villa Sub-Exchange, W. M. Moseley of Fort Pierce representing Fort Pierce Sub-Exchange, H. C. Allan of Oak Hill representing North Indian River, J. B. Prevatt of Tavares representing Lake County, A. R. Updike of Lake Wales representing Lake Region, Jack N. Strong of Vero Beach representing Indian River, A. V. Saurman of Clearwater representing North Pinellas, Phil C. Peters of Winter Garden representing Orange County, Ford W. Moody of Palm Harbor representing Pinellas, Armer C. Johnson of Mount Dora representing Plymouth, D. A. Hunt of Lake Wales representing Polk County, Charles G. Metcalf of Avon Park representing Scenic Sub-Exchange, and George B. Aycrigg of Winter Haven representing Winter Haven Sub-Exchange.

Officers for the 1948-49 season were W. C. Van Clief, president; Phil C. Peters, first vice-president; A. V. Saurman, second vice-president; H. C. Allan, third vice-president; John L. Olson, fourth vice-president; Counts Johnson, secretary; James Samson, assistant secretary; and S. L. Looney, treasurer.

Department heads elected for the 1948-49 season were C. C. Commander, general manager; Fred S. Johnston, general sales manager; S. L. Looney, comptroller; Counts Johnson, general counsel; Carlisle Kyle, advertising manager; E. E. Gaughan, traffic manager; and Frank J. Poitras, canning division manager.

CHAPTER 33

1948-1950

THE FLORIDA citrus industry opened the 1948-49 season in the depths of an industrial depression which, in the opinion of Exchange officials, had been mostly self-imposed. Both Nature and good fortune, however, were disposed to look kindly on Florida during the two-season period from 1948 to 1950. Gradual higher prices to growers were in evidence almost from the beginning of the period, and the slowly rising tide of good fortune had led to a comfortable profit margin by the close of the 1948-49 season and to undeniable prosperity by the close of the 1949-50 season.

The higher price trend of the 1948-49 season was caused principally by killing freezes in California and Texas that drastically reduced the national supply of citrus. Florida, basking in the mildest winter on record, virtually monopolized the nation's citrus market after the historic California freeze of January 3 to 11 and also the Texas freeze on the last two days of the same month. Alone in the market the Florida industry probably surprised the trade as much as itself by resisting to a great extent the temptation to flood the markets under these favorable conditions. A voluntary volume-prorate plan of controlled shipments, introduced by the Florida Citrus Commission at the request of growers and shippers, was apparently subscribed to by a sufficient majority of shippers to hold supply volume in check during this period. With regard to the prorate plan, it should be noted that this cooperative action among shippers and growers was both unusual and commendable in Florida, particularly in the light of a severe drought condition that threatened excessive droppage in mid-season varieties in the central and southern parts of the citrus belt.

Also contributing to the rising tide of fortune for the Florida citrus grower was the high quality of Florida citrus during the 1948-49 season. With exceptional quality existing in all Florida fruit, and the doubtful quality of the freeze-damaged fruit from California and Texas, it seemed that Floridians had, for the moment, every possible advantage for a good season. Not to be overlooked in the Florida industrial picture was the sensational increase in the demand for frozen orange concentrate and the resulting rush of the trade to increase supplies of this product.

It is interesting to note that the Florida Citrus Exchange completed an agreement with the Snow Crop firm during the 1948-49 season for the sale of two million boxes of oranges to Snow Crop for concentrating purposes. While the sale itself was extremely important to the economy of the industry, an agreement between the two firms to establish a minimum price to assure the cost of production was an unprecedented move that acted as a tonic to the industry. Prior to this agreement, the average delivered-in prices being paid by canners during the season peaked at $0.65 per box for oranges and $0.35 per box for grapefruit. By February 14, 1949, following the completion of the contract between Snow Crop and the Florida Citrus Exchange, these prices had jumped to $1.20 for oranges and $0.60 for grapefruit. A further increase placed these prices at $1.95 and $0.75 by the close of the month of March.

By April 1, 1949, the industry's first major battle between fresh and concentrate interests was shaping up into an unrealistic war for the remaining supply of oranges. Canners were offering as much as a $2.00 per box, delivered in, for oranges, and grapefruit prices were ranging from $0.75 to $0.90 per box. At the same time the f.o.b. price for fresh oranges was up to $3.25. Late in April the f.o.b. price for both oranges and grapefruit approached the $4.00 mark and by May 16 small-sized oranges were bringing $5.00 per box.

Canneries had packed fewer oranges but more grapefruit and tangerines during the 1948-49 season than during the previous season, and the growth of the frozen concentrate industry had been phenomenal. While three concentrate plants had packed a total of 1,700,000 gallons of concentrate during the 1947-48 season, the 1948-49 season saw ten plants packing an estimated 10,000,000 gallons. The total volume of the Florida citrus crop for the 1948-49 season was just under 93,000,000 boxes, a figure somewhat under the original estimate owing to the droppage of fruit mentioned earlier in this chapter.

The Exchange considered the season a definite financial success insofar as its returns to Exchange members were concerned. A summary of the season by Sales Manager Fred S. Johnston was included in the annual report for the 1948-49 season:

> Fruit marketed by the Exchange brought higher prices this season than at any other time since the drastic decline in prices shortly after the close of the war. Compared with the two previous seasons, this year has been a definite financial success for growers, who had been in the depths of a depression while virtually all other farmers were enjoying the greatest period of prosperity in the history of the nation.
>
> Higher citrus prices were the result of killing freezes in California and Texas, a successful state-wide voluntary volume prorate introduced by the Florida Citrus Commission, the fine quality of Florida citrus this season, the increased purchases of citrus for frozen concentrate, and the below-normal carry-over of canned citrus stock.

Under its canning division, the Exchange continued to increase activities in all trade areas. Appointment of Frank J. Poitras as sales manager of the canning division was made on August 1, 1949, and sales of SEALD SWEET juices were made in thirty-four states and Canada during the summer period.

The appointment of Guy E. Howerton as manager of the Exchange and Supply Cooperative during the early part of the season had followed the general reorganization of the Supply Company, and by the close of the 1948-49 season the cooperative had reported nearly a 500 per cent increase in business over the preceding season. Thus the Exchange, along with the rest of the industry, completed the 1948-49 season with the feeling that the Florida citrus grower's lot was on the mend. The American public had money and wanted citrus, and the housewife was willing to pay a reasonable price for good-quality fruit. Per capita consumption of citrus had increased from twenty-six pounds to eighty-seven pounds in the past two decades, while consumption of most competitive fruits had declined during the same period. The Exchange was convinced that the American housewife would continue to buy an increasing amount of citrus, especially as the public was further educated by the advertising of health values of citrus and citrus products.

There was no change in the board, its officers, or in the staff executives of the Exchange at the close of the 1948-49 season, which also saw the introduction of the familiar, all-encompassing Citrus Code, generally supported by the Exchange, and its acceptance by the state legislature.

Then the memorable 1949-50 season began. The outlook for Florida was considered good, and growers looked hopefully to the new Florida Citrus Mutual for the stabilization of prices at least above the cost of production. Florida citrus growers reaped the richest harvest in the history of the industry during the 1949-50 season, and the continued prosperity was due in great part to the phenomenal growth of the frozen concentrate business. Popularly called "The Cinderella Product" by the nation's leading writers, frozen concentrate had come of age, and its coming would control the destiny of the industry for years to come.

Other factors were present, however, to make the 1949-50 season both progressive and prosperous. California suffered its second straight season of devastating freezes, Texas production was curtailed because of the freezes of the past season, Florida production was less than had been anticipated because of a hurricane, and the nation's payroll reached its highest point in the history of the country.

Out-of-state capital began flowing into the Florida citrus industry in almost unlimited proportions as investors became entranced with citrus and most particularly with the frozen concentrate industry. Modern concentrate plants were constructed in almost every section of the citrus belt, and existing facilities were extensively remodeled and enlarged. Additional millions of dollars were spent to expand mechanical facilities for the production and transportation of the big Florida citrus crop, which had originally been estimated at 58,800,000 boxes of oranges, 24,000,000 boxes of grapefruit, and 5,000,000 boxes of tangerines. In addition to this production one of the heaviest multiple blooms in the memory of growers indicated the probability of a record-breaking crop for the 1950-51 season.

As for the Exchange during the 1949-50 season, this analysis appears in the annual report of that season:

> Although the volume of fresh-fruit shipments declined about 30 per cent this season, because of the increased percentage of fruit moving to the concentrate plants, there will always be a market for fresh fruit. Consumers are demanding higher quality, however, and more careful attention must be given to the grading, packing, and keeping quality of fresh citrus. The new maturity law, in effect for the first time this season, helped to insure better quality. Although there were some protests against the law, none of these protests were upheld in the courts. Some growers complained that the higher maturity law would prohibit the marketing of certain varieties of citrus, but there were no varieties that could not eventually be marketed in one of the three outlets: fresh fruit, frozen concentrate, or single-strength canning.

The Florida Citrus Exchange did not move fruit in volume into markets during the 1949-50 season until late in October because of the higher maturity standards of the new Citrus Code, and because of a later-than-usual bloom. Opening prices were high and the trade was anxious, so that by November 5 oranges at auction were bringing an average price above $5.00 per box.

Just prior to the Christmas season the market, demoralized somewhat, reached $3.32 by the close of November, whereupon Florida Citrus Mutual took control of the situation by establishing minimum prices of $2.35 f.o.b. for packed oranges and $1.10 for cannery grades. The Exchange gave credit to this action for the immediate stability it gave to the market.

The market continued to rise during the early part of the year 1950, as California suffered its freezes. Frozen concentrators began to purchase in heavy volume, and history was made as the market steadily advanced to previously unheard of prices. By the end of the early and mid-season orange deal, concentrators were paying $3.50, delivered, for oranges, and the Exchange reported on March 11 that a total of 1,353 cars of oranges had brought an average of $6.37 at the ten major auctions.

As retail prices soared, sales began to drop noticeably and the Exchange reported that buyers were withholding orders by mid-April in anticipation of lower prices. Florida Citrus Mutual again stepped into the picture, and on April 20 established minimum prices for oranges at $3.25 to $3.75 and $2.50 for oranges delivered to canning or concentrate plants. This action apparently brought renewed stability to the market, with buyers placing orders with confidence that these prices would be maintained. The balance of the season was completed without serious threat to the over-all economy.

As the season closed, the industry could take stock of its finest period. Florida had shipped 26 per cent less fresh citrus in all varieties, but favorable prices had compensated for this loss of tonnage. The total of all varieties processed during the season showed only a slight increase owing to a great reduction in the grapefruit pack. Processors had used only 58 per cent of the total movement, and of the processed oranges, 50 per cent had been used by frozen concentrators. It is interesting to note that the frozen concentrators had utilized 105 per cent more citrus during the 1949-50 season than during the 1948-49 season. Railroads had handled 46 per cent of the total of all shipments, while truck carriers had handled 47 per cent and ships had accounted for the balance. This compared with the previous season of 60 per cent by rail, 40 per cent by truck, and virtually no transport by ship. Total production for the season had been just short of 88,000,000 boxes, with a total of 58,500,000 boxes of oranges, 24,200,000 boxes of grapefruit, and 5,000,000 boxes of tangerines.

Because of the years of experience already gained by the Florida Citrus Exchange, there was a note of caution injected amid the industry's unparalleled optimism at the close of the 1949-50 season. Exchange officials pointed to several factors that would affect the coming season in almost direct proportion to the control exercised by the industry. Extremely high prices, warned the Exchange, had lulled many growers into a feeling of security that was not justified by economic factors. This caution was issued by the Exchange to its membership at the close of the season:

> While next season should be profitable to the grower, there is no reason to believe that prices will be fantastically high or that the grower will receive more than a reasonable, modest profit on his investment. In fact, there are grave possibilities of an oversupply next season if the big crop is not marketed in orderly fashion.
>
> Early estimates indicate a record-breaking crop for Florida during 1950-51.
>
> California production, off 20 per cent this season, will be almost back to normal. Texas, likewise, is recovering at a rapid rate and

will market considerably more citrus next season. The total national citrus crop will approach or break all records, which means this big crop obviously must be marketed in an orderly fashion.

If the ever-increasing production of citrus is going to be marketed, it will be necessary to utilize to the utmost capacity all avenues of distribution: fresh, single-strength, and frozen concentrate.

Frozen concentrate plants in Florida utilized 13,706,000 boxes of oranges through May 13 and can be expected to utilize even more during the next season. The extent of this increase will depend upon such factors as whether the quality is maintained, the rate at which food stores throughout the nation can be equipped with frozen food compartments, and the competition of other fruit juices.

A recent survey indicates frozen concentrate is meeting with excellent acceptance when the retail price does not exceed $0.29 for a six-ounce can. This survey also reveals that frozen concentrate is more competitive to single-strength canned citrus than to fresh citrus. Frozen concentrate will increase the total consumption of citrus products only if the quality of this popular product is maintained.

The consumption of fresh citrus has more than doubled during each of the last several decades as the nation has become more conscious of the health qualities and delicious taste of citrus juices. There will always be a demand for fresh citrus regardless of how much frozen concentrate is sold. The consumer, however, will demand fresh citrus that tastes as good as it looks and will promptly turn to frozen concentrate if the fresh fruit available is not of comparable or better quality.

The total consumption of citrus can be expected to continue to increase, especially considering that the four biggest concentrating facilities will spend considerably more money advertising Florida citrus than the growers themselves in Florida have ever spent in a single season.

It was primarily an increase in consumption that enabled citrus to be the only agricultural commodity that received satisfactory prices this season, except those commodities subsidized by the federal government.

Although not mentioned in the above analysis of the outlook for the 1950-51 season, the Exchange also looked with alarm at the threat of foreign citrus moving into the nation's markets because of good demand and high prices. This threat also worried the Exchange because such movement carried the danger of entry of major plant pests into this country. That this was a real danger is indicated in a record of congressional action in Washington by congressmen from Florida, California, Texas, and Arizona demanding a federal embargo against truck shipments of fruit from Mexico as protection from the citrus blackfly. In this regard, the annual report of the Exchange for the 1949-50 season carries this notation:

Another foreign threat is the Oriental fruit fly. The rapidity with which this fly spread in Hawaii, after being brought to that Island from Saipan during the war, indicates that it could obtain a reasonably strong foothold in almost any part of the United States.

Senator Holland recently opposed any move to ship citrus fruit from Cuba into the United States through Florida ports. He wrote Secretary of Agriculture Brannan that such importation "would immediately endanger Florida's multi-million dollar vegetable and citrus industry because of the risk of entry of major plant pests on such fruit."

Despite the increasing threat of foreign pests, the United States government considered a plan to relax to some extent its inspection of passengers' baggage from abroad. The Florida Citrus Exchange Board of Directors resolved "that it is unalterably opposed to the proposed plan of spot or percentage inspection of passengers' baggage from abroad, a system that would provide for the unrestricted entry from foreign countries of thousands of units of plant material, a considerable portion of which is likely to be infected with insects and diseases capable of causing serious economic injury to the horticultural and agricultural investments of the state and nation."

While there is evidence that the Exchange's warning in this latter business of plant pests was given consideration, the issue was unfortunately never settled to any satisfactory extent, although the infestation of the fruit fly in the mid-1950's was eventually to emphasize the soundness of this warning.

CHAPTER 34

1950-1952

DURING the life of the Florida Citrus Exchange its policies and philosophy had seldom changed rapidly except in the face of emergencies that compelled hasty action. Most clearly defined changes had occurred slowly and deliberately and were usually preceded by changes on the board of directors or on the executive staff. The two-season period from 1950 to 1952 saw several important changes in the membership of the board and in the top executive positions. The inexorable passage of time, as well as fast-moving developments within the industry, were now combining to close out another era for both the Exchange and the industry.

There can be little doubt that the timely development of the frozen concentrate industry was in itself a tremendous factor in the period immediately following the close of the 1949-50 season. The rapidly rising acceptance of Florida Citrus Mutual as a stabilizing agency most certainly had its effect on the entire industry, and the movement of the leadership of the industry from one generation to the next was bringing in its wake certain progressive changes that were to set the pattern of the industry as it is known in present times.

While the Florida Citrus Exchange entered the 1950-51 season with but one new addition to its board, Tom B. Swann of Winter Haven representing the Florence Sub-Exchange, the season was to see considerable change on both board and the staff before the conclusion of the season. The appointment of A. V. Saurman as general manager of Florida Citrus Mutual eventually moved him to resign on September 1, 1950, as an officer of the board and as a director of the Exchange. This vacancy as a representative of the North Pinellas Sub-Exchange was filled by the appointment of O. J. Harvey, himself a former district manager in the Exchange system. On January 25, 1951, the death of S. L. Looney terminated thirty years of service to the Exchange organization, the last seventeen years of which he had served as the treasurer-comptroller as well as the executive vice-president of the Growers Loan and Guaranty Company. The board then unanimously named James Samson, an employee of the Exchange since 1926 and in more recent years the assistant secretary of the organization, to the positions formerly held by Looney. On March 22, 1951, W. C. Van Clief tendered his resignation as president of the board and as a special director of the Exchange. Named to succeed him as president was J. B. Prevatt, a member of the board representing the Lake County Sub-Exchange.

As the board completed the 1950-51 season it was composed of A. R. Updike of Lake Wales, I. J. Pemberton of Jacksonville, Tom B. Swann of Winter Haven, W. M. Moseley of Fort Pierce, Jack N. Strong of Vero Beach, Charles G. Metcalf of Avon Park, J. B. Prevatt of Tavares, John L. Olson of Dundee, H. C. Allan of Oak Hill, O. J. Harvey of Tampa, P. C. Peters of Winter Garden, F. W. Moody of Palm Harbor, Armer C. Johnson of Mount Dora, D. A. Hunt of Lake Wales, E. G. Todd of Avon Park, and George B. Aycrigg of Winter Haven. President and chairman of the board was J. B. Prevatt; P. C. Peters was first vice-president; H. C. Allan, second vice-president; John L. Olson, third vice-president; and F. W. Moody, fourth vice-president.

One of the most far-reaching changes in the Exchange system during the 1950-51 season occurred on April 1, 1951, with the retirement of C. C. Commander as general manager. Commander, who had held the top executive position for twenty-six consecutive years, by both temperament and talent had been one of the recognized leaders of the industry during its development from the mid-1920's. The board of directors of the Florida Citrus Exchange had realized for many months that its general manager was reaching retirement age and had appointed a special committee for the selection of a replacement for the retiring veteran. This committee, after many weeks of detailed study, brought back a unanimous recommendation for the appointment of John T. Lesley to the position. Lesley, one of the state's leading cooperative management experts, accepted the appointment and took up the duties of general manager on April 1, 1951. Lesley came to the Exchange from the Haines City Citrus Growers Association, where he had served as manager for the previous eight years. Relatively young in years, he had served on various citrus committees of national scope, and he represented to the board an ideal combination of youth and experience.

Meanwhile, Commander had agreed to continue serving the Exchange wherever possible in an advisory capacity. Thus, the "Little General," as Commander was known throughout the industry, completed his tenure of active service to the Exchange and to the industry. Forcefully dominant, extremely shrewd, and undeniably dedicated, Commander had served the Exchange during both its most difficult and its most successful years. It is doubtful that any other single individual within the industry during these times could have left a more indelible mark than did Charley Commander.

At the close of the 1950-51 season, staff executives of the Exchange were, in addition to Lesley, Fred S. Johnston, sales manager; James Samson, comptroller; E. D. Dow, traffic manager; Counts Johnson, general counsel; and Carlisle Kyle, advertising manager.

The season itself was a year of war jitters as the conflict in Korea developed. The Florida citrus industry enjoyed what the Exchange described as a moderately profitable season. Statistics of the season reveal that the expected record-breaking production did materialize into a crop of ninety-six million boxes which was marketed in a relatively orderly manner.

A review of the season seems to indicate that most of the peaks and valleys in the price structure of previous years were replaced by a stable market that neither dipped into red ink nor soared to impractical heights. Again, considerable credit for this stability must certainly have gone to Florida Citrus Mutual. The Florida Citrus Commission, with an enlarged advertising fund as a result of a larger crop, bought more advertising space and time which, in turn, continued to bring additional increases in the consumption of citrus fruits. Lowest average prices for the season occurred during November, 1950, when fresh fruit returned $3.41 per box for oranges, while grapefruit at the same time was bringing $4.15. Canners prices during a part of the season skidded as low as $0.40 per box for grapefruit and $1.27 for oranges delivered in. In both instances, Mutual's price minimums served to stem the low price trends.

Texas, during the season, had suffered a severe freeze in January. Experts compared the damage to Florida's "Big Freeze" of 1894, and the Exchange's Ford W. Moody, visiting in the Texas area at the time, wired:

> Freeze damage terrible—entire valley. Every grove damaged, many killed outright. Ninety-five per cent bearing trees killed to banks. Approximately twenty per cent were not banked—these all dead. Authorities here estimate total loss four million trees all ages. Practical fruitmen say it will be a miracle if production is a million boxes this season. With normal temperatures it will be three years before some commercial production—five to seven years before normal commercial production.

The Exchange closed out its 1950-51 season for most practical purposes with its annual meeting on June 7, 1951. Two changes on the board of directors occurred, with C. G. Wilhoit replacing H. C. Allan—after his death in January, 1951—as the representative of the North Indian River Sub-Exchange, and J. P. Ellis replacing D. A. Hunt as the representative of the Polk County Sub-Exchange. Staff positions remained the same except for the addition of Marion J. Young as the manager of the frozen concentrate division and the appointment of Walter J. Page as public relations director.

The 1951-52 season produced the largest and one of the highest quality citrus crops in the history of the industry. This crop was marketed at ridiculously low prices in the opinion of the Florida Citrus Exchange. Total volume for the season reached nearly 111,000,000 boxes with nearly 2,000,000 boxes of the rapidly increasing Temple variety of orange included in that total. By mid-October, Florida Citrus Mutual, which had received considerable credit the previous season for maintaining a satisfactory price structure, had signed 210 of the state's 228 fresh-fruit shippers to Mutual contracts. There was a rather heavy carry-over of processed citrus, especially canned grapefruit juice, but an intensive summer campaign by the Commission and others had increased consumption. Consumers were drinking frozen orange concentrate at a record-breaking pace that reached 727,000 gallons during the week of September 29, 1951. Before the beginning of November, Mutual had established $1.03 per box delivered as the minimum price for the lowest grade of oranges which could be accepted by processors. At the same time it established $2.15 per box as the minimum price for the lowest grade and least-desirable size at which fresh-fruit shippers could sell oranges at the fo.b. level. In early December the f.o.b. market for oranges was firm at $2.50 to $2.15, depending on sizes. Grapefruit were selling from $2.50 upward for Duncans, $2.50 upward on Marsh, from $4.50 upward on Pinks, and tangerines were bringing from $2.00 upward. Heavy shipments in December were brought to a halt by the longest shipping holiday ever attempted by the industry, eleven days beginning December 21.

Processors were taking oranges at the rate of 1,250,000 boxes per week by mid-January, 1952, and retail sales of frozen concentrate continued at a high level. During the first week in March, Florida passed the halfway mark in harvesting its largest-to-date volume production. The Valencia movement was increasing rapidly, and heavy shipments resulted in a weakening of the markets. Then Mutual established the first in a series of prorates to send prices upward slightly. The three major outlets for Florida oranges—fresh, canned, and frozen—were using fruit at such a fast pace during March that, despite an estimated increase in the size of the crop, there were fewer than three million boxes more of fruit remaining to harvest at the close of March than were remaining at the same time in the previous season, in spite of the great production of the 1951-52 season.

A price war developed during May between producers of frozen concentrate which brought prices on the finished product down to a point that aroused alarm and some criticism from Florida growers. However, by the close of May there were indications of a trend upward in both processed and fresh fruit and the season concluded without serious dips in the general price structure established over the entire period.

It is interesting to note that the Exchange had been campaigning to increase its f.o.b. business while maintaining its auction outlets at the highest possible point for several seasons past. The trend toward f.o.b. purchases that began with the nation's swing to the supermarket type grocery had continued to develop to such proportions that it had become of vital importance to all fresh-fruit marketing agencies in the state. For the season, Exchange Sales Manager Fred S. Johnston reported an increase in the cooperative's f.o.b. sales totaling 39 per cent.

It might be well to consider at the close of this chapter some facts with regard to the history of the industry to this point. While the days of returns at less-than-production cost for growers were still remembered, and although growers of grapefruit were still facing extensive difficulty, the economic climate of the Florida citrus industry had been slowly stabilizing, at least somewhat above the critical stages, for entire seasons at a time.

Apparent to most of the industry was the fact that production would continue to increase at an annual rate of about 10 per cent per year under normal weather conditions for some time. While the increase in consumption had been almost phenomenal, the increase in production could be expected to overshadow increased utilization at some date in the distant future unless industry could step up the rate of increased consumption. More by-products of citrus were needed, and interest in the development of these by-products was lagging. Hard-hitting advertising in sufficient quantity would certainly be necessary to keep demand in relation to increasing supply, quality would need to be improved, and further Florida product identity was needed.

The Florida Citrus Exchange, gathered up in the concentrate movement, had completed negotiations with the Snow Crop firm during the season for the processing of Exchange fruit. In addition, the newly activated concentrate division was already engaged in the sale of SEALD SWEET frozen concentrate across the nation. Years of promotion of the SEALD SWEET brand in fresh and canned form was paying off handsomely in the organization of the frozen concentrate division of the Exchange's sales department.

At the conclusion of the 1951-52 season, General Manager Lesley made the following report to the board:

We are near the close of what appears to be the largest year in Exchange history. Our volume is roughly one million boxes greater than the volume of last season.

We were determined to increase our share of the fresh-fruit market and, in order to do this, we re-established our own inspection service so that we could assure our customers of a more uniform grade and pack.

We created a dealer service department, inaugurated a monthly trade bulletin, added to our sales department, and secured excellent cooperation from the individual associations in filling very complicated orders. It is difficult to single out just what one thing did the most for us, but we do know that collectively these innovations resulted in substantial increases in our f.o.b. sales this season.

We have established a market for SEALD SWEET frozen concentrate and will be capable of selling all frozen concentrate available to us.

Even though the Exchange's membership has less need for a home for their cannery fruit than possibly any other single group in the state, the Exchange thoroughly investigated the possibility of purchase of the Clinton Food facilities, and when this was deemed advisable we entered into an arrangement with Clinton to assure the opening of their Florida plants, as it was felt by all that if these plants were not opened there would be three or four million boxes of fruit in the state without a home, a situation that could have adverse effect on the market.

Although we feel the Exchange has had a very successful year, we are not satisfied. We firmly believe that this industry must have greater cooperation in marketing and distribution if we are ever to level out the sharp and sometimes unnecessary decline in price.

With regard to the near purchase of the Clinton Food facilities by the Exchange, it is interesting to note that the purchase had at one time drawn the full approval of the board, but floundered on one phase of the technical details involving the assumption of control by the Exchange. Even after the collapse of negotiations about concentrate facilities, the Exchange was successful in making possible the continued operation of the Clinton facilities during the season.

In February, 1952, the Exchange had been named the successful bidder on 761,695 gallons of hot-pack concentrate for the National School Lunch Program. This marked the first attempt by the Exchange to obtain a government contract for concentrate. In cooperation with Clinton Foods, Inc., the cooperative revitalized a sagging market in the process of fulfilling this contract. Exchange growers received a total of $691,102 for 585,680 boxes of fruit packed under government order.

The board of directors conducted its annual meeting on June 5, 1952. There were three changes in the board for the 1952-53 season. Alfred A. McKethan of Brooksville was seated as the representative of the North Pinellas Sub-Exchange; J. P. Garber was named as a director representing the Lake Byrd Sub-Exchange, replacing Charles G. Metcalf; and O. J. Harvey was now seated as a representative of the Elfers Sub-Exchange. Subsequently, following the death of W. M. Moseley of Fort Pierce on March 6, 1952, H. H. Willis, Sr., of Fort Pierce was seated on the board.

Appointed to staff executive positions for the 1952-53 season were John T. Lesley, general manager; Fred S. Johnston, sales manager; James Samson, treasurer-comptroller; H. S. Weber, traffic manager; Counts Johnson, general counsel; and Walter Page, director of advertising and public relations. Frank J. Poitras continued as sales manager of the canning division, and Charles W. Metcalf as sales manager of the frozen concentrate division.

President and chairman of the board was J. B. Prevatt, first vice-president was Phil C. Peters, second vice-president was C. G. Wilhoit, third vice-president was John L. Olson, and fourth vice-president was Ford W. Moody.

CHAPTER 35

1952-1954

THIS HISTORY has heretofore endeavored to extract the most significant developments of each season within the Florida Citrus Exchange and within the industry wherever feasible. This has been possible to some extent in the light of the effect of these developments on the now known progressive history of the industry. With the close of the 1951-52 season we now approach current times—times that deal with individuals currently prominent in the Florida citrus scene. Unlike the details of earlier, more clearly defined periods since 1909, the history of the past eight years is still in the making.

To attempt to extract the most significant developments of these more recent years within the industry would be difficult indeed, for the developments of these years have not yet approached their historical conclusion. Accordingly, this history will from this point on concentrate almost exclusively on developments within the Exchange organization, leaving the current progress of the industry to the exacting columns of statistics that will perhaps form the foundation for other histories of future years.

During the 1952-53 season, in view of all factors affecting supply and demand, the Exchange considered the industry's marketing of its 102,000,000-box crop as fairly satisfactory. However, there were some matters of concern for the future. General Manager John T. Lesley had this to say in a report to the board at the end of the season:

From 1943-44 to 1951-52, Florida's orange crop increased from 46,200,000 boxes to 78,600,000, an average of slightly more than 4,000,000 boxes per year. The problem then is to keep step with this tremendous increase. Florida's relative prosperity will be judged by how well the job is done.

To offset even greater production to come, Florida growers, shippers, and processors must fight to retain every box now being sold as fresh fruit and single-strength canned juice. National consumption of frozen concentrated citrus juices on a wider scale must be developed further.

Money earmarked for research is always money well spent. Industry appropriations for research into new fields designed primarily to bring the best possible tasting juice to the greatest mass of consumers at the lowest possible cost consistent with reasonable returns to growers is a "must" if Florida's citrus empire is to survive and prosper.

Meanwhile, the Exchange had continued its drive for increased fresh f.o.b. sales and, at the close of the 1952-53 season, had increased this category of business by nearly 80 per cent during the period from 1951 to 1953. Exchange officials were, however, concerned about the seriousness of "spreading decline," and had joined the industry in appealing to Washington for additional funds to halt this tree disease which was becoming of significant importance to growers in some sections of the state.

Prior to the close of the 1952-53 season, the organization had combined its canning division and its concentrate division into a single organization known as the Florida Citrus Products Exchange. A new and lucrative sales outlet had been found for single-strength juices and sections in the export markets of Belgium and France, and plans for sales to South America were being made for the following season.

Production and sale of SEALD SWEET concentrate had reached slightly over one million boxes of citrus during the season, including special packs of lime concentrate and lemonade made possible through the membership of the Florida Tropical Fruit Growers Association in the Florida Citrus Products Exchange.

The annual meeting of the Florida Citrus Exchange was held on June 4, 1953. Seated at this meeting, or subsequently seated on the board, was John C. Updike of Lake Wales representing Alcoma Sub-Exchange, I. J. Pemberton of Jacksonville representing Clark Sub-Exchange, O. J. Harvey of Tampa representing Elfers Sub-Exchange, R. K. Cooper of Winter Haven representing Florence Sub-Exchange, H. H. Willis, Sr., of Fort Pierce representing Fort Pierce Sub-Exchange, Jack N. Strong of Vero Beach representing Indian River Sub-Exchange, J. P. Garber of Avon Park representing Lake Byrd Sub-Exchange, G. B. Hurlburt of Mount Dora representing Lake County Sub-Exchange, John L. Olson of Dundee representing Lake Region Sub-Exchange, J. B. Prevatt of Tavares representing Lake Region Packing Association, C. G. Wilhoit of Wabasso representing North Pinellas Sub-Exchange, P. C. Peters of Winter Garden representing Orange County Sub-Exchange, F. W. Moody of Palm Harbor representing Pinellas Sub-Exchange, Armer C. Johnson of Mount Dora representing Plymouth Sub-Exchange, J. P. Ellis of Bartow representing Polk County Sub-Exchange, E. G. Todd of Avon Park representing Scenic Sub-Exchange, and George B. Aycrigg of Winter Haven representing Winter Haven Sub-Exchange.

The 1952-53 slate of officers of the board was carried over to the 1953-54 season and staff positions, except for the appointment of H. S. Weber as traffic manager following E. D. Dow's retirement, were also unchanged.

The 1953-54 season saw the Florida Citrus Exchange launching the largest advertising-merchandising program on fresh fruit that it had undertaken in many years. Under a plan devised by the general manager, strong merchandising efforts were supported by a concentrated advertising campaign designed to reach millions of housewives across the nation. Both trade and consumer advertising were carried throughout the season. The program was financed through an assessment levied for advertising purposes. An arrangement during the season with the Florida Lychee Growers Association resulted in the membership of that organization in the Exchange for the marketing of the relatively unknown lychee crop during the summer break following the close of the 1953-54 citrus season.

A special committee, headed by G. B. Hurlburt of Mount Dora, was appointed during the season to consider the possibility of the redesignation of the Exchange's sales department under a new name that would incorporate SEALD SWEET as a brand-corporate identity. After extensive study and several meetings concerning this move, the board accepted the name "SEALD SWEET Sales, Inc.," as the designation for its sales department. Final approval of the name was voted following the close of the 1953-54 season for implementation at the beginning of the 1954-55 season in September.

With regard to the season itself, Florida produced more than 90,000,000 boxes of citrus. Processors broke all past records in production of frozen orange concentrate, turning out an estimated 67,000,000 gallons. Single-strength producers exceeded their previous year's pack by an estimated 1,000,000 gallons and packed 1,600,000 more cases of blended juice than they had packed in the 1952-53 season.

Average auction prices for Florida's fresh oranges during the season were up about five cents per box over the previous season, while grapefruit prices were nearly 31 cents below the seasonal average for 1952-53. The grapefruit crop, largest on record, reached above 42,000,000 boxes for the season and amounted to an estimated 87 per cent of the nation's total supply. Florida sent 38, 230 cars of grapefruit to market at a seasonal average of $2.91 for Duncans, and $3.88 for Seedless. While prices were from 31 to 41 cents below the prior season, the industry was generally agreed that the significant increase in production was largely responsible.

The revival of the European export markets for Florida citrus was now reaching significant proportions with a volume of one million boxes of Valencias alone shipped to Europe during the season. Meanwhile, SEALD SWEET concentrate was distributed widely in forty-two states and four Canadian provinces during the season, utilizing more than 1,270,000 boxes of Exchange growers' fruit.

The Florida Citrus Exchange at the close of the 1953-54 season continued to predict higher production. It was not alone. The Continental Can Company, at the conclusion of a comprehensive survey, had estimated that total citrus production in Florida would approximate 150,000,000 boxes by the 1956-57 season, barring a crop disaster. Its survey concluded that this production would be composed of 104,000,000 boxes of oranges, and 46,000,000 boxes of grapefruit and tangerines. This production would be utilized by 26,000,000 boxes fresh, 12,000,000 boxes single-strength, 63,-000,000 boxes in frozen concentrate, and 2,500,000 boxes in other processed forms.

Pointing to these figures, General Manager John T. Lesley in his annual report to the board at the close of the 1953-54 season included these remarks:

These figures indicate the increased production will have to be utilized by frozen concentrate, but they also point up the fact that it is absolutely necessary to maintain the single-strength and fresh-fruit markets. It is for this reason that the Exchange has put so much effort in developing and strengthening its sales and merchandising program. We are convinced that, properly handled, our fresh-fruit market will not only continue to use a large quantity of fruit but will also serve as a good indicator for the proper pricing of our fruit to processors.

The immediate trouble spot in our citrus picture is grapefruit. The present USDA estimate indicates we have produced more than 40 million boxes. In spite of increasing our commodity advertising expenditure to a total of 6 cents per box, we have been unable to obtain satisfactory prices.

With greater competition facing us from increasing Texas production, it is vital that we get our house in order by doing those things we all know we should do.

1. We should revise our minimum standards for maturity and quality on both fresh and canned grapefruit to assure the consumer a better-tasting product.
2. We should push vigorously for both state and federal regulations that will permit control of volume and diversion of surpluses.
3. The industry's advertising and merchandising program should be strengthened by more brand advertising.

Thus the Florida Citrus Exchange completed the season of 1953-54, and its hopes for the future as indicated in the general manager's recommendations virtually duplicated the philosophy of the Exchange in years gone by.

The annual meeting of the Exchange was held on June 3, 1954. Directors seated at this meeting and their affiliations were John C. Updike of Lake Wales representing Alcoma, I. J. Pemberton of Jacksonville representing Clark, O. J. Harvey of Tampa representing Elfers, R. K. Cooper of Winter Haven representing Florence Villa, H. H. Willis, Sr., of Fort Pierce representing Fort Pierce, Jack N. Strong of Vero Beach representing Indian River, J. P. Garber of Avon Park representing Lake Byrd, G. B. Hurlburt of Mount Dora representing Lake County, J. B. Prevatt of Tavares representing Lake Region Packing Association, John L. Olson of Dundee representing Lake Region, C. G. Wilhoit of Vero Beach representing North Indian River, Alfred A. McKethan of Brooksville representing North Pinellas, Phil C. Peters of Winter Garden representing Orange County, F. W. Moody of Palm Harbor representing Pinellas, Armer C. Johnson of Mount Dora representing Plymouth, J. P. Ellis of Bartow representing Polk County, E. G. Todd of Avon Park representing Scenic, and George B. Aycrigg of Winter Haven representing Winter Haven. Aycrigg was later replaced on the board by E. S. Horton of the Winter Haven Sub-Exchange.

Officers of the Board carried over from the previous season were J. B. Prevatt, president and chairman of the board; P. C. Peters, first vice-president; John L. Olson, second vice-president; C. G. Wilhoit, third vice-president; F. W. Moody, fourth vice-president; James Samson, treasurer-comptroller; and Counts Johnson, secretary.

Staff executives appointed for the 1954-55 season were John T. Lesley, general manager; Fred S. Johnston, sales manager; Charles W. Metcalf, concentrate division sales manager; Frank J. Poitras, canning division sales manager; H. S. Weber, traffic manager; Counts Johnson, legal counsel; J. Samson, comptroller; and Walter J. Page, director of merchandising and public relations.

The continued improvement of the export markets in Europe resulted in the employment of Howard N. Baron during the summer months for the purpose of establishing an export division for the sale abroad of Exchange products both fresh and canned.

CHAPTER 36

1954-1956

THE FLORIDA citrus industry produced slightly more than 130,000,000 boxes of citrus during the 1954-55 season, this production being made up of 88,400,000 boxes of oranges, 34,800,000 boxes of grapefruit, 5,100,000 boxes of tangerines, and 2,400,000 boxes of Temple oranges. In its annual report for the season, the Florida Citrus Exchange included these comments on the season:

The start of this season followed the pattern of all previous seasons in that oranges started off at exceedingly high prices and quickly dropped to much lower levels.

When the government crop estimate came out on October 11, 1954, showing 96 million boxes of oranges, the market eased off considerably. As we moved further into the season, heavy shipments brought about lower prices, with prices at Christmas ranging between $2.25 and $2.50 per box f.o.b.

These low prices continued through January, with competition very keen and movement heavy. This was brought about by the fact that the high government estimate and the fear of a freeze kept fruit moving at these low, competitive prices without regard to a more orderly marketing program.

The latter part of January the f.o.b. market moved up to $2.40 and remained steady until mid-February, when it advanced to $2.75 after the week-end freeze scare of February 11th and 12th. This, together with the reduction of two million boxes of oranges from the government estimate of February 10th, reduced shipments and increased prices at f.o.b. to $3.00 to $3.25 where they remained generally for the rest of the season.

For grapefruit, like oranges, the season began early at high prices and soon slumped to mediocre or low prices. The state produced more tangerines this year than last, but they were much smaller sizes, presenting an extremely difficult marketing problem. However, more tangerines were shipped this year at slightly better prices through the aid of the Florida Citrus Commission and the Florida Tangerine Cooperative.

The exchange, under its new export division, had considerably increased its shipments of fresh and processed products to Europe over past seasons, and Florida oranges were being delivered in excellent condition abroad.

The Exchange's canning division had marketed SEALD SWEET single-strength juices in many United States and Canadian markets as well as in export markets. An important addition to the SEALD SWEET line of canned products during the season was the introduction of six-ounce cans of sugar-added orange and grapefruit juices for the beverage trade. A combination of circumstances during the season had forced single-strength orange juice prices upward—a decrease in the orange crop estimate and greatly increased purchases of oranges by concentrators. Grapefruit juice prices, on the other hand, had declined, with inventories considerably greater than in the past season.

Again, excellent progress was reported by the concentrate division of the Exchange, which reported at the close of the season that SEALD SWEET frozen juices had been sold in forty-four states and four Canadian provinces. A newly introduced frozen unsweetened line of Indian River grapefruit juice in 6- and 32-ounce sizes was promising extremely good consumer acceptance by the season's close.

The Growers Loan and Guaranty Company had completed its thirty-eighth year of service to the Florida Citrus Exchange on April 30, 1955. During these thirty-eight years the Guaranty Company had lent nearly eighty-eight million dollars to Exchange members and their growers, and the close relationship between the financial company and Exchange associations and their members continued to work to the advantage of the Exchange system as a whole. It is interesting to note, in passing, that another Exchange affiliated organization, the Exchange Supply and Service Cooperative, reported at the close of the season that its sales had more than doubled in the seven-year period since its reorganization in 1938.

The general manager of the Exchange, John T. Lesley, concluded the organization's 1954-55 season with this message to the board:

Growers and shippers of fresh Florida citrus face a real challenge during the next two years. A date with destiny could obliterate many or cure all, depending upon how we meet the crisis.

The fresh-fruit segment of the industry must be revitalized and strengthened if we are to maintain a marketing balance between fresh and processed, a balance very vital to the welfare of the grower. This projected exigency appeared to be hastened this season when the first crop estimate of October indicated an oversupply of oranges. However, subsequent reductions in estimates curtailed supplies and brought them into line with what is calculated to be normal usage.

Production is on the increase in almost all growing areas of Florida, Texas, California, and Louisiana. Texas alone, which this year produced 4 million boxes, is expected to market about 12.5 million within the next two years.

Moreover, a very disappointing trend in Florida continues to be amplified. Fresh-fruit shipments are being drastically reduced and more and more Florida fruit being diverted into cans. Fresh-fruit markets must be expanded in order to keep the industry healthy and the grower prosperous. How can this be done?

1. *Better internal quality:* To hold fresh-fruit customers shippers must offer fruit of better internal quality, at least on a par with standards for frozen concentrated juices. Only premium fruit of the very highest quality should be offered to fresh-fruit customers.

2. *Better external quality:* Eye appeal is the housewife's measuring stick for buying her citrus fruits. External appearance must be of the best in order to compete for bin or display space in modern markets.

These two requisites, together with the timing and handling of shipments so that Florida citrus reaches the consumer at the peak of condition, will do much toward solving the major share of our troubles. Add to this, strong brand advertising, sales promotion, and merchandising support, and marketing ills should dissipate appreciably.

The annual meeting of the Florida Citrus Exchange at the close of the 1954-55 season was held on June 9, 1955, with only slight changes in the board of directors resulting. C. G. Wilhoit, a board member who for several years represented North Indian River, was now seated as a representative of the Graves Brothers Company. The vacancy thus existing was filled with the election of J. C. Flake of Mims. D. A. Hunt, a member of the board representing the Polk County Sub-Exchange for many years, was returned to the board as a representative of the Hunt Brothers Cooperative of Lake Wales. Both the Graves and Hunt Brothers' organizations were, so far as the organizations were concerned, new members of the Florida Citrus Exchange.

The appointment of Traffic Manager Paul C. Sarrett to replace retiring H. S. Weber was the only change in the staff executive roster as a result of the 1954-55 annual meeting, although activities of the 1955-56 season were to effect several significant developments.

While, for the most part, routine annual meeting affairs were conducted at the meeting of June 9, 1955, the Exchange made its first official move toward the establishment of prepackaging facilities in the New York terminal market area. The plan, as discussed at this time by the board, was to include the shipment of bulk fruit from Florida to the Exchange's packaging facility in New York, where the fruit would be prepackaged in consumer-size packages for retail distribution. The beginning of the 1955-56 season was to witness the end of the Florida Citrus Products Exchange and once again the return of the Florida Citrus Exchange to the exclusive marketing of fresh fruit.

For many months the dual role of the large cooperative in both fresh and processed marketing activities had been under discussion by the board. It was, perhaps, inevitable that board members representing individual associations of the Exchange would eventually find it impossible to agree on the conduct of the business with regard to processed products.

It must be remembered that the Exchange, as a matter of policy, had long since agreed that each association should arrange for the sale of its cannery fruit in any manner most beneficial to its growers. With this in mind, it seems logical that geographical location as well as other important factors should create a situation in which some Exchange associations were supplying fruit to Exchange-affiliated processing plants, while others were not. This division of interest eventually culminated in the withdrawal of the Exchange from its activities in the marketing of all processed and canned products. The Florida Citrus Products Exchange thus became the nucleus for the Plymouth Citrus Products Cooperative, and the working force was transferred from the Exchange to Plymouth, where SEALD SWEET products are still manufactured under an agreement spelled out at the beginning of the 1955-56 season. This move marked the conclusion of the second venture of the Exchange into the canning business, the first endeavor having been terminated at the peak of the depression in the mid-1930's. From this time until the present, the Florida Citrus Exchange has functioned solely in the interest of marketing the fresh-fruit production of its members.

The 1955-56 season itself was considered by the Exchange to have been the most favorable since the conclusion of World War II. Good quality grapefruit brought reasonably good prices as the industry concentrated on the grapefruit returns problem. Early and mid-season orange prices were stabilized near the beginning of the season when processors purchased considerable volumes at $1.25 on the tree. This apparently placed a sort of minimum, and higher prices were placed on these varieties for the balance of the shipping season. On February 29, 1956, Valencia orange prices were ranging from $3.00 to $3.75 per box f.o.b., and were advanced another 25 cents by mid-April. Top prices were reached on April 30, with another advance of 25 cents.

With regard to the prepackaging plans of the Exchange mentioned briefly earlier in this chapter, the trial-run phase of the program was conducted in Philadelphia during December, 1955. From this experimental operation, utilizing already established commercial prepackaging facilities, the Exchange hoped to set the pattern for its contemplated self-owned facility in New York. The Philadelphia experiment was also considered as a sort of sounding board through which to consider additional prepackaging operations in other cities such as Chicago and Detroit. The initial phase of the prepackaging operation was considered successful by the board, and plans were continued to purchase a prepacking operation in the New York area.

On December 8, 1955, J. B. Prevatt submitted to the board his resignation as president of the Florida Citrus Exchange following an illness that would curtail his personal activities to such an extent as to make it impossible for him to serve in this office. After accepting this resignation, the board elected Phil C. Peters of Winter Garden to serve out the remaining term. Elected to serve as vice-presidents were John L. Olson, first vice-president; C. G. Wilhoit, second vice-president; F. W. Moody, third vice-president; and G. B. Hurlburt, fourth vice-president.

Thus, into the top position of leadership in the Exchange system came the highly respected Orange County grower and shipper. Peters, who had served intermittently on the board since 1918, brought with him the prestige and knowledge of years of experience in citrus, and a significant background in cooperative marketing.

A wide gap between the end of the mid-season crop and the beginning of the Valencia shipping season brought additional stability to the markets even though concentrate plants were not in operation to draw off excess supplies. During the months of March and April extremely dry weather prevailed and many of the citrus areas were threatened by the drought. This, as always, forced the movement of many crops sooner than anticipated, but growers, holding the line on prices, came through the threat at no disadvantage. Exports to Europe from Florida of nearly one and a half million boxes of fresh citrus again helped to maintain the price structure. The Exchange, with its fully implemented export division, was, of course, a large and important export utilization factor. There can be no doubt that this had been another year when shipments of fruit through fresh channels were hard pressed to compete with processing prices. In its annual report for the season, however, the Exchange expressed confidence that its members had received seasonal fresh-fruit averages comparing favorably with prices for processed fruit.

Much of the Exchange's favorable position in the fresh-fruit market was undoubtedly a result of the continued merchandising program. While the promotional activities of the previous year had been curtailed somewhat as a result of the transfer of its processed interests, the Exchange continued to promote its SEALD SWEET brands in all major distribution areas. The 1955-56 season, although profitable for growers throughout the state, was viewed by the Exchange as one of the contradictions stemming from the government's crop estimate. Although a reduced volume of oranges had gone into fresh marketing channels, there was a considerable increase in the shipment of grapefruit. Florida Citrus Exchange shipments, reflecting the state-wide reduction, were down, but f.o.b.

volume, both percentagewise and in total volume, was the largest in history. In view of this, the Exchange was optimistic over the fresh-fruit marketing outlook and was already at work on experimental tests designed to assure the consumer of better citrus fruits.

There were, however, certain problems on the horizon at the close of the 1955-56 season. It will be remembered that the Exchange had joined the industry several seasons prior in expressing alarm over the serious threat of the "slow decline." When the burrowing nematode became definitely identified as the cause, it had also been determined that the only control was to push out the infected trees and fumigate the soil. The State Plant Board, along with the United States Department of Agriculture, had already been charged with this responsibility. By the close of the season some 1,145 acres had been treated in this manner.

Perhaps the most serious problem of all was one discovered on April 21, 1956. On a dooryard tree in the Miami area had been found the larvae of the dreaded Mediterranean fruit fly. By the end of April the fruit fly had been found as far north as Lake Wales, dashing all hopes that it could be contained in the noncommercial areas in South Florida. While the origin of the initial entry of the fruit fly could not be traced, of course, a resolution directed to the Department of Agriculture by the Exchange some years prior with regard to more detailed inspection of baggage at Florida ports of entry to the United States had gone unheeded. What the Exchange feared had come to pass, and the industry could now look to a long and expensive process of elimination of the fruit pest. Destruction of the fruit fly in the late 1920's had cost $7,500,000, and that Florida had been able to completely annihilate the pest had been a modern miracle. At any rate, it soon became obvious that the Florida citrus industry would now become involved in a tremendous effort to rid itself of the fly, and by May 15, 1956, the estimated cost of the eradication process had already reached $25,000,000 with an immediate outlay of $9,000,000 facing the industry.

Nevertheless, the Exchange looked to the 1956-57 season with much optimism. Canned inventories were low and, with the devastating Spanish freeze during the preceding winter, the possibility of substantial export shipment was good. This, plus the estimation of a lighter crop, made prospects favorable for the board as it concluded its business for this season.

CHAPTER 37

1956-1958

THE BOARD OF DIRECTORS of the Florida Citrus Exchange for the 1956-57 season was seated at the annual meeting held on June 6, 1956. The board was composed of John C. Updike, Alcoma Association, Inc.; I. J. Pemberton of W. H. Clark Fruit Company; A. V. Saurman of Clearwater Growers Association; O. J. Harvey of Elfers Citrus Growers Association; R. K. Cooper of Florence Citrus Growers Association; H. H. Willis of Fort Pierce Citrus Growers Association; D. A. Hunt of Hunt Brothers Cooperative; A. N. Strong of Indian River Citrus Sub-Exchange; G. B. Hurlburt of Lake County Citrus Sub-Exchange; John L. Olson of Lake Region Citrus Sub-Exchange; J. B. Prevatt of Lake Region Packing Association; John C. Flake of North Indian River Citrus Sub-Exchange; Alfred A. McKethan of North Pinellas Citrus Sub-Exchange; Phil C. Peters of Orange County Citrus Sub-Exchange; F. W. Moody of Pinellas Citrus Sub-Exchange; Armer C. Johnson of Plymouth Citrus Growers Association; J. P. Ellis of Polk County Citrus Sub-Exchange; and E. S. Horton of Winter Haven Citrus Growers Association.

Officers of the board for the 1956-57 season were Phil C. Peters, president and chairman of the board; John L. Olson, first vice-president; C. G. Wilhoit, second vice-president; F. W. Moody, third vice-president; and G. B. Hurlburt, fourth vice-president. Counts Johnson was elected secretary and James Samson was elected treasurer-comptroller.

Executive appointments for the 1956-57 season were John T. Lesley, general manager; Fred S. Johnston, general sales manager; Counts Johnson, general counsel; H. N. Baron, export sales manager; and Paul C. Sarrett, traffic manager.

The consumer bag division, discussed in prior chapters, was thoroughly implemented as an Exchange service during the 1956-57 season. This division was headed by Edward H. Wales, who had been appointed as the sales manager of this phase of the Exchange's sales operation. By the midway point in the season, the Exchange was employing commercial bagging organizations in Philadelphia, Boston, Pittsburgh, Baltimore, and Chicago. The Exchange's own bagging facility at Kearney, New Jersey, was placed in operation before the close of the season.

The 1956-57 season was to see the infestation of the fruit fly reach its maximum proportion, but all-out cooperation by the industry, the State Plant Board, and the United States Department of Agriculture served to develop packing-plant fumigation systems that precluded imposition of the difficult embargoes of the infestation of the 1920's. General Manager John T. Lesley, remarking on the fruit-fly problem in June, 1956, had said: "The extra expense of treating fruit is regrettable, but the fly is something we can live with—we can still market good sound fruit. We are in a better position than we were in the 1920's because of better fumigating equipment in packinghouses and better insecticides for spraying groves." While the industry was undoubtedly faced with a long period of expense and inconvenience, the initial shock of the discovery of the pest soon gave way to a feeling of confidence in the ability of the industry to eradicate successfully the fly while profitably marketing the major portion of the crop.

The 1956-57 season was another record breaker by volume with 93,000,000 boxes of oranges, 37,400,000 boxes of grapefruit, 4,800,000 boxes of tangerines, and slightly more than 3,000,000 boxes of tangelos and Temples. The season began auspiciously for both fresh-fruit growers and shippers. Efforts of the Growers Administrative and Shippers Advisory committees to keep high-quality levels imposed on early shipments of oranges and grapefruit received state-wide support from the industry. This attitude was hailed by the Exchange and much of the industry as indicative of the great strides toward effective cooperation that had been made by the industry in just a few short years. That this was true, there can be little doubt. The most confining early-season grade and size regulations in the history of the industry were imposed by the two control committees with hardly a dissenting voice from the industry. Cooperative awareness of the marketing importance of high-quality fruit continued throughout the season, bringing to almost every segment of the industry a renewed dedication to the principles of cooperative marketing so long upheld by the Exchange as those necessary for continued prosperity within the industry.

The 1956-57 season, as it progressed, saw independent growers and cash-buying processors again locked in a price battle. The situation obviously followed a trend that had been building for several years, and one that was causing considerable concern within the industry. Although growers had usually managed to hold their own in past years, strong factors in the 1956-57 season seemed to align themselves on the side of the buyers. The first of these factors was, of course, the record-breaking production coupled to an extremely high juice yield of the season. The second was that, even though retail processed prices had been substantially reduced from comparable prices of the previous season, sales had increased by only about 1 per cent and high inventories were threatening. Third, heavy and well-timed rains throughout the citrus belt had broken a three-year drought, and all elements were present for a heavy crop in the 1957-58 season.

As the trade began to show reaction to these factors, concentrators began edging down the selling price of finished concentrate, and prices to growers started downward. While this situation eventually righted itself, it brought to light certain advantages of cooperative marketing held highly important by the Florida Citrus Exchange. In the opinion of the Exchange, this situation pointed to the uncertain and often vulnerable position of the so-called independent grower. It is rather obvious that the independent producer is in an excellent position, during the years of short supply, to obtain the highest price for his fruit; in years of abundant crops, his position is worsened substantially from that of a cooperative producer. As General Manager John T. Lesley perceptively noted in the annual report for the 1956-57 season: "The grower who has protected his position by having an assured home for his fruit, both cannery and fresh, is in good position to face the increased production that, barring disaster, the future will bring."

Thus, the Florida Citrus Exchange entered the 1957-58 season with confidence and optimism. At its annual meeting on June 6, 1957, the board accepted the resignation of John L. Olson because of illness, and seated Joe E. Keefe of Dundee to fill the vacancy. A. V. Saurman of Clearwater Growers Association likewise resigned from the board, his vacancy being filled by E. S. Beeland of the Clearwater organization. Officers for the 1957-58 season were Phil C. Peters, president and chairman of the board; C. G. Wilhoit, first vice-president; F. W. Moody, second vice-president; G. B. Hurlburt, third vice-president; and Joe E. Keefe, fourth vice-president.

It should be noted that the membership of the board and the officers in the Exchange have continued through to the 1959-60 season with but one exception. J. B. Prevatt, following the withdrawal of the Lake Region Packing Association from the Exchange at the close of the 1957-58 season, resigned from the board at that time. The executive staff also remained unchanged during this time, except for the transfer of William G. Strickland, formerly manager of the Indian River Citrus Sub-Exchange, to the Tampa headquarters as assistant to the general manager.

As has been indicated, the Exchange entered the 1957-58 season with considerable optimism. Events of the prior season with its record production, tighter controls, higher quality, and cooperative effort seemed to substantiate this feeling of optimism. In addition, a considerable amount of satisfaction stemmed from the fact that the fruit fly eradication program had been sufficiently successful that the end of the fly menace was in sight. But the 1957-58 season had no sooner settled down for the long pull than cold weather began to threaten. After a series of "scares," the "big freeze" blew frigidly into the Sunshine State during December to bring the most critical freezing conditions experienced by Florida in sixty years. The December freeze was followed by another in January. Stunned, the Florida citrus industry stood almost helpless as Nature wrecked an estimated 40 per cent of its expected crop of 142,000,000 boxes. Many young groves were completely killed and hundreds of acres of mature groves were so severely injured that fruit on them was completely lost for the season.

In the opinion of many, the industry was held together during these trying times by the immediate action of the Florida Citrus Commission. The Commission moved promptly to prevent panic utilization of freeze-damaged fruit in finished processed products, embargoed movement of fresh fruit out of the state for a short time, and voiced its determination to maintain high-quality standards in spite of the severe adversities facing the industry. These actions gained the immediate confidence of the trade and, in this respect, the industry was to learn gradually that it had at long last weathered a severe freeze without the historic market slumps caused by the shipment of low-quality, freeze-damaged fruit. Convinced that Florida was policing its quality rigidly, the major portion of the trade responded with reasonable confidence, and the structure of the markets in all channels was held intact.

It is interesting to note that the industry eventually received higher gross returns from its drastically curtailed production than it probably would have received had the entire crop been marketed. Prices skyrocketed while the industry discovered almost in amazement that the nation's consumers would pay more for their citrus than had ever before been realized. It is probable that this reaction of consumers was the first conclusive proof of the value of the commodity advertising program of the industry. Citrus had obviously been accepted as a necessary part of the nation's diet, and consumers were paying the world's highest peacetime price for their citrus products. The Florida Citrus Exchange managed to serve its members well during this hectic season. While the state average for the fresh-fruit shipments continued to slide downward during the season, shipments of Exchange fruit had shown a substantial increase. President Phil C. Peters had this to say at the close of the season:

This has been indeed a critical year for our entire citrus industry. It is gratifying to know, however, that all of our members of the Florida Citrus Exchange have for the most part fared better than the average during the freeze. The fast-moving and efficient marketing and sales organization coupled with our lines of communications have provided our members with every advantage pricewise in a fast-changing and rapidly moving market. It is logical to assume that out of the past winter will come research in many fields to provide greater cold tolerance plus better quality and production.

As the season of the "big freeze" came to a close, the citrus industry was forced to the conclusion that those who had not been totally dispossessed by the weather had completed an unusually good marketing season. This almost unbelievable situation was to usher in the greatest period of confidence ever known by the Florida citrus industry. Though the most feared of all adversities, prolonged freezes, had been undergone by the Florida citrus grower, the season had closed out in such excellent condition that even the veterans in the industry were hard put to believe the return figures.

Actually, production for the season was at its lowest point since the 1952-53 season and the certainty of a short crop in 1957-58 further maintained high price levels. Perhaps the only casualties among the growers who sold their fruit immediately following the freezes were the independent growers who sold to cash buyers for concentrate at panic prices. Concentrators, taking a historic gamble that will not soon be forgotten, doubled work forces and maintained peak operations throughout the freeze period. Under Florida Citrus Commission regulations, concentrators using freeze-damaged fruit were required to hold concentrate for a specific period of time, then subject it to strict quality tests before placing it in consumer containers for shipment. As the concentrate industry went into twenty-four-hour-a-day operations following the freezes, no one could know for certain that his entire pack would not be outlawed for consumer shipment.

That the long gamble paid off is now legend within the industry, but the panic sales by independent growers to cash-buying concentrators immediately following the freezes cost these growers millions of dollars, and added substantially to the high profits received by the concentrators as a reward for this gamble. It is a matter of record that grower-members of cooperative concentrate firms received many millions of dollars over prices received by independent growers who were forced to dispose of their fruit on a day-to-day basis during the disastrous days of the freeze period.

With renewed confidence in its own capability to meet disaster, the Florida citrus industry entered into what many believe to be the greatest era of cooperation. Certainly greater cooperative efforts of the industry during the fruit-fly infestation and the freezes, as well as the result of this cooperation in the matter of price levels, substantiated the theory that most of the industry's problems could be greatly reduced by cooperation at every level in Florida citrus. Speaking on this newly found spirit of cooperation, General Manager Homer Hooks of the Florida Citrus Commission commented:

There is a spirit of cooperation and of working for the common good of the industry prevailing in Florida that veterans in the industry have not seen for many years.

This is not accidental atmosphere. It is the outgrowth of a new maturity in our industry, a realization that the welfare of the growers, shippers, canners, concentrators, and all those who service them are inevitably intertwined.

Whether the industry has, in reality, reached a new maturity is a question that remains veiled in the uncertainty of the future. But those who have watched its development over the years believe that a new spirit has been found that will provide a strong foundation for progress.

Now this history approaches the final chapter. To be sure, there remains a great abundance of problems yet to be faced. The shifting economy of the nation, as well as that of the world, places much of the future on an elastic platform that cannot be made firm. The industry's marriage to the whims of Nature also serves to place its seasonal future forever beyond human direction. But, working in harmony, the industry may now contemplate a future that can be devoid of the extreme ups and downs of years gone by. Long-range plans may be laid with more assurance, and research into the several sciences of citrus already promises new and vitally important utilization development. Beyond this, the Florida Citrus Exchange takes confidence in the trend of the industry to resist all efforts to lift final control of its destiny from the capable hands of thousands of growers who form the solid foundation on which this empire is constructed.

Note

CHAPTER 38

1958-1959

AS HAD BEEN predicted, the 1958-59 season was one of short supply and high prices. Total production of oranges reached eighty-six million boxes, and production of grapefruit amounted to thirty-five million boxes. On the whole, there had never been a more profitable year for growers, but the emphasis was on concentrate. Lesser volumes of fruit and the resultant high prices critically curtailed movement in fresh-fruit channels as processors battled for oranges to fill depleted inventories. It is a matter of record that prices demanded for oranges eventually reached such proportions that fresh-fruit distributors were forced into an unprecedented situation. For the first time in the history of Florida citrus, the sale of fresh grapefruit at the Florida Citrus Exchange exceeded the sale of fresh oranges. This single factor—the emphasis on varieties other than oranges—was responsible for a relatively successful 1958-59 marketing season at the Exchange. Total volume was about equal to the year preceding.

Holding its annual meeting on June 4th, 1959, the board of directors heard General Manager John T. Lesley outline the trends of the 1958-59 season:

With regard to the specific activities of our Sales Department, this report cannot hope to overcome the obvious. It has not been one of our better seasons because of the many developments with which you are all familiar. California produced one of the largest and finest crops of oranges ever produced in that state, to compete with Florida citrus which was in short supply at less than its highest quality and holding capacity.

With speculative concentrators' prices at an all-time high, it was necessary for us constantly to raise our prices from the very beginning of the Valencia season. As our price picture elevated, many markets turned to the lower-priced California oranges. Florida, because of this turn of events, shipped a lower percentage of Valencias in fresh fruit than ever before in history.

While the matter was only briefly mentioned during the annual meeting of the Exchange, the impending retirement of General Sales Manager Fred S. Johnston could be expected to result in the selection of a new general sales manager early in the 1959-60 season. Eventually named to replace Johnston was Assistant Sales Manager Donald M. Lins, who would assume the duties of top sales executive on January 6, 1960. A graduate of Cornell University with considerable background in agriculture, Lins would move easily into the complexity of the coming season.

Rather typical of the complexities of the citrus industry is the fact that while fresh-fruit interests struggled for supplies during the 1958-59 season, the industry as a whole was enjoying its highest crop valuation in history. Total packinghouse door-level valuation of the season's production was eventually placed at nearly $348,000,000, as compared to the 1957-58 season valuation of $273,000,000, itself a record-breaker. Both years had been difficult for fresh-fruit interests. By the close of the 1958-59 season, citrus leaders throughout the state were showing considerable concern over the decline in movement of fresh citrus. To those involved in the fresh-fruit channels of trade, the problem seemed to be clearly etched in terms of supply and demand.

Processors with the capability of blending and holding could control their distribution of products manufactured from highly priced, poorly conditioned fruit. With expected low supplies creating high processor demand, prices moved even the most dedicated fresh-fruit shippers into the concentrate plants. Through all this one apparently undeniable trend was being established: Fresh-fruit shippers had become a sort of buffing element within the industry. In times of short supply, fresh-fruit interests are all but forgotten in the general movement stampede toward the canning plants where storing potential and distribution control can return higher-than-average profits to growers. Speculative summer sales by processors willing to gamble—sometimes disastrously to themselves and the grower—further complicated the matter. Conversely, in times of high production, processors return pricing structure to levels that provide growers with the opportunity to place good quality, premium fruit into fresh-fruit channels at favorable profits. Historically, in normal years, fresh fruit has always returned more to the grower than he has received from the canner. It is because of this circumstance that the industry, at the close of the 1958-59 season, looked with considerable misgiving at the difficulties that the fresh-fruit industry had encountered during the two preceding years. As troubles mounted for fresh-fruit shippers, their marketing avenues were beginning to dissolve for lack of merchandise.

Thus, the board of directors of the Florida Citrus Exchange concluded the 1958-59 season with some concern for the future. The board, unchanged in membership from the preceding season, called for a situational evaluation from its top executive, General Manager John T. Lesley. His evaluation of the future of the fresh-fruit industry in Florida on June 4, 1959, is quoted in full:

> I should like to preface these remarks with a note of optimism for both the Florida Citrus Exchange and the fresh-fruit industry in general. The 1959-60 season will, in my opinion, see a considerable increase in movement of fruit in the fresh channels of trade.
>
> With a return to near-normal production of all varieties, and in view of the processor inventories that have accumulated, it appears likely that fresh-fruit shippers will be called upon throughout the 1959-60 season to dispose of a larger quantity of fruit than during the past few years.
>
> In addition to this supply and demand factor, it is my observation that the industry itself is becoming genuinely concerned about the declining proportion of the total production now being utilized in fresh-fruit channels.
>
> The logical conclusion that can be drawn from these factors is that fresh-fruit shippers will, in all probability, experience an active season throughout the 1959-60 period. It is possible that the coming season can be the best fresh-fruit shipping season in the past several years, from a volume standpoint.
>
> I believe, too, that returns to growers will continue to reflect favorable profits, although it is difficult to predict any sort of price range until we have some idea as to the rate of depletion of processor inventories. It appears certain, however, that the bulging warehouse inventories of concentrate at this time will pose a difficult

price problem at least during the first part of the new season, particularly in view of the fact that these inventories have been accumulated from extremely high-priced raw products.

While the foregoing remarks depict a rather favorable outlook for the 1959-60 season, I must at the same time direct your attention to the general future outlook for all fresh-fruit interests in the state. As I have said, I believe the coming season will be a good one for fresh fruit. Beyond next season, the difficulties loom formidable and imposing.

I would like to discuss some of these difficulties with you, and I believe it of paramount importance that we begin this discussion with an analysis of our most important single focal point—the market place.

There have been drastic changes in our markets. Consider, if you will, the rise of the supermarket within a few short years, and the increased concentration of buying power that this has caused.

As our retail outlets have shifted from the corner grocery store to the modern supermarket, our methods of supplying these outlets have become outdated in the transition. This change has been rapid, and almost frightening to suppliers at times. Last year for example, 68 per cent of all stores were relatively small outlets which did less than 8 per cent of the total grocery business. In other words, 203,000 small grocery stores processed only 8 per cent of the total business, while 28,000 supermarkets sold 67 per cent of all grocery items moved through regular food outlets.

Another indication of the speed of this transition from small to large stores is the fact that as recently as 1952 supermarkets were doing only 43 per cent of the grocery business as compared to today's approximately 70 per cent.

Accompanying this increase in the size of the stores doing the major part of the grocery business has been an increase in the number, size, and buying power of grocery chains. And, although we have always had a few large grocery chains, the increase in the number of large chains has certain significance.

But perhaps of greater significance is the emergence of small independent chains consisting of eleven stores or fewer. These relatively small chains of supermarkets still do most of the grocery business—slightly over 60 per cent.

And, in this area, we face the latest in the long list of changes that have presented problems to the fresh-fruit industry. Something new has been added in recent years. A large majority of the so-called independent chains are now organized into voluntary chains through which to take advantage of volume buying.

Thus, within a few short years, virtually our entire market structure has changed. Terminal markets have declined almost as rapidly as supermarkets have taken over. Direct purchasing by a few large buyers has become the most important factor in the sale of fresh Florida citrus. Wholesalers and brokers are disappearing from the scene at an alarming rate, and each short-supply season accelerates this rate of disappearance.

As we have moved into this era of concentrated buying power, the face and make-up of our buyer have changed drastically. Today's high-volume, fast-moving fruit buyer is, first of all, a businessman. He thinks and acts like a businessman, and he is well paid to show profit on every item in his province.

He demands a continuous supply of high-quality merchandise in the package, size, and price range that is attractive to the consumer. He wants uniformity, and he seeks always to minimize both the amount and the expense of the effort that must be expended between shipper and consumer.

In addition to these factors, the very nature of self-service supermarkets requires help from the supplier in the matter of advertising and merchandising at the consumer level. And, in today's modern grocery store, few items are moved that have not been pre-sold by brand or commodity promotion to the extent that the consumer is familiar with them before she buys.

This, then, is a picture of our markets in general as we prepare to do business with them today.

Now let us consider the development of our fresh-fruit industry to cope with the changing market.

Our production has increased. Our trees now produce more fruit of commercial quality than was thought possible a few short years ago. We have, in many instances, overlooked quality in an effort to cash in on high-yield production, yet our quality in general has increased gradually and continually. We have sufficient production to meet demand, except during natural disasters that occur infrequently. We are unhampered by government crop controls, and we have invented manufacturing processes that adequately handle that part of our production which cannot be moved into fresh channels. We have even developed by-products that are made from the refuse and pulp of the canning plants, and this in itself has grown into big business.

It would seem, therefore, that with adequate production and high-volume demand, we have established a perfect climate in which to do business. This is true, with one highly important and disturbing exception. We have made little or no progress in the intermediate phase of our business which places the supplies of the producer into the hands of the retailer. In plain language, our fresh-fruit sales agencies are little changed in this period of great change, and somehow we seem to resist the idea that progress should and must be made in our methods of selling.

It is true that we have converted a portion of our sales effort to the so-called f.o.b. market. But we have done even this in a somewhat reluctant manner.

What else have we done? Nothing, to speak of—and most of our fresh-fruit problems today stem from this simple fact. While our markets have changed and grown, and our production has increased, we continue to sell our product in much the same way it was sold fifty years ago.

We are meeting the concentration of buying power with the same disorganized, nervous, and spasmodic sales methods that have caused us trouble ever since the industry faced the problem of marketing a total of six million boxes back in 1909. But, until the great changes in the markets of recent years, along with the development of the concentrate industry, organization could be disregarded without the serious repercussions of today. In the past, when three hundred separate sales agencies could play the whole field of terminals, brokers, wholesalers, and peddlers in wild abandonment of sound marketing procedures, the sources of utilization were numerous at any rate.

Today, nearly two hundred separate sales agencies are still attempting to squeeze the same methods of distribution into a rapidly narrowing funnel of distribution outlets. The results of this tragic circumstance are apparent in the statistics of the fresh-fruit industry.

Volume buyers are forced to go from agency to agency in search of adequate supplies, and there is seldom assurance that today's supplier can be depended upon to fill next week's order. Prices are erratic as small shippers vie for a part of the volume trade, and buyers must simply hope that the prices they have paid will be approximately the same as those paid by their competitors. Quality, too, must suffer as a consequence of this system of marketing.

Thus, the buyer of fresh Florida citrus is given, at best, few of the factors he considers so important in the conduct of his high-volume business. His supply is seldom dependable, his quality is an up-and-down thing, and his price is a gamble pure and simple.

He has virtually no brand promotion assistance, and is encouraged only by the fact that the Florida citrus industry has a state-administered commodity advertising and merchandising program that is supported by industry mostly because it derives its money from taxes on the entire citrus enterprise.

I think that you will agree that this is a disturbing portrayal of our selling methods. It is, in my opinion, also the answer to our declining fresh-fruit market.

While the solution to this problem is apparent enough, the method by which to accomplish the solution is evasive. Somehow we must manage to reduce drastically the number of fresh-fruit sales agencies in operation today. We must do this without decreasing the tonnage available for fresh-fruit utilization, and we must do it without disturbing the control of the industry, which remains where it should be, in the hands of the Florida grower.

We must concentrate our selling power in a parallel with the concentration of buying power. In doing so, we place vastly increased supply potential in the hands of the remaining sales agencies. The cost of selling along with the cost of packaging can be decreased as greater volume moves through fewer selling points.

Advertising and merchandising of select industry-wide brands distributed nationally can be accomplished at a surprisingly low cost per unit to supplement the commodity program already in existence, and uniformity of package and product can be more nearly attained than ever before in the history of the industry.

The Florida citrus industry stands now at the crossroads of destiny, in my opinion. If we are to regain the ground lost in the decline of the fresh-fruit business, we must set about to reshape our business to the pattern of our changing markets.

If, on the other hand, we ignore the decline of the fresh-fruit industry in the brilliant light of the processor's progress, we must be prepared for the consequence of a single outlet business hinged exclusively on the dictates of the canner.

I believe, however, that most of us in the industry today know full well the importance of maintaining our fresh-fruit channel of utilization. I am optimistic in the belief that the Florida citrus industry has, indeed, entered a new era of cooperation that will set the stage for faster and better development in all areas of our endeavor.

This cooperation, when applied to such fields as nutritional research, marketing, packaging, new product research, and business administration, seems likely to open new and expanded vistas all across the broad future of the industry.

The fresh-fruit industry is in difficulty to be sure. But the Florida citrus industry has only to survey the far greater difficulties of other agriculture producers, in industries that have allowed their fresh-product utilization to wither away, to find the will and determination to retune our business to the times.

As for the Florida Citrus Exchange, fifty years in Florida citrus have given to us a position of prominence and leadership. I am confident of our united desire to direct this leadership toward the development of new ideas and new approaches to the wonderful challenge of keeping in step with this industry of ours—which, through its peculiar blessings, has progressed more often than not.

Index